Steve Jones is Emeritus Professor of Genetics at University College London. He delivered the BBC Reith Lectures in 1991, appears frequently on radio and television and is a regular columnist for the *Daily Telegraph*.

By the same author

The Language of the Genes
In the Blood
Almost Like a Whale
Y: The Descent of Men
The Single Helix
Coral: A Pessimist in Paradise
Darwin's Island

THE
SERPENT'S
PROMISE

The Bible Retold as Science

STEVE JONES

ABACUS

First published in Great Britain in 2013 by Little, Brown
This paperback edition published in 2014 by Abacus

3 5 7 9 10 8 6 4

Typeset in Bembo by M Rules
Printed and bound in Great Britain by
Clays Ltd, St Ives plc

Papers used by Abacus are from well-managed forests
and other responsible sources.

MIX
Paper from
responsible sources
FSC® C104740

Abacus
An imprint of
Little, Brown Book Group
100 Victoria Embankment
London EC4Y 0DY

An Hachette UK Company
www.hachette.co.uk

www.littlebrown.co.uk

To the memory of my great-grandfather
William Morgan, for forty years preacher at
Capel-γ-Garn, Bow Street, Ceredigion

CONTENTS

ACKNOWLEDGEMENTS

I thank Sam Berry, Neil Bradman, David Colquhoun, Jerry Coyne, David Ellis, Iain Hunt, Andrew Huxley, Nick Lane, Hugh Montgomery, Linda Partridge, Norma Percy, Peter Robinson, Andres Ruiz-Linares and Tim Whiting for their comments: any remaining errors are, needless to say, my own.

What, it will be Questiond, When the Sun rises do you not see a round Disk of fire somewhat like a Guinea O no no I see an Innumerable company of the Heavenly host crying Holy Holy Holy is the Lord God Almighty.

William Blake,
Notebook on a Vision of the Last Judgement (1810)

The one duty we owe to history is to rewrite it.

Oscar Wilde,
The Critic as Artist (1881)

PREFACE TO THE PAPERBACK

Every author should take the advice of Saint Luke that '. . . as ye would that men should do to you, do ye also to them likewise'; in other words, 'if you can't take it, don't dish it out'. I usually follow his counsel, but I was nevertheless rather startled by some of the responses to the original version of this volume. Perhaps I should add a few words to this paperback printing in an attempt to clarify what the book is and – more important – what it is not.

The Serpent's Promise has, as subtitle, *The Bible Retold as Science*. The reaction by some among the faithful to that latter phrase has been that of a Calvinist to a consecrated wafer: whatever the merits or otherwise of the titbit on offer, believers cannot swallow it: 'Cue heavenly choir of white-coated lab dwellers . . . worshipping the Great Beard of Darwin, which looks remarkably like that of God', or, if you prefer, 'a re-cranking of the Darwinian barrel organ – accompanied by the monkey of New Atheism, of course, as it screeches petulantly at religion' (and that one was so apposite that it makes it to the back cover of the present edition).

In truth what I aim to do is to interpret the Good Book not from the point of view of an atheist (which I happen to be) but to read it as an attempt to make sense of the physical world and hence as an ancestor of today's science. What have we found out in the past two millennia and more about the origin of life, about inborn fate, about the visions of the prophets or even about religion itself? Can science ever hope to explain why we do 'unto others' by acting in a way that diminishes our own prospects while benefiting those of our fellows?

The Serpent's Promise tries to answer those questions and others like them. It is not an attack on religion but a journey into Dawkins Canyon; the yawning and largely unexplored chasm between the faithful and their opposites which is, in truth, filled with a variety of remarkable creatures. Adam and Eve are, as genetics makes clear, the real ancestors of every man and every woman today and are not mere models of the loss of innocence that comes with carnal knowledge. Leviticus' obsession with purity is often interpreted as a sermon on moral spotlessness, but it contains a hidden truth, for the book was written as great cities grew and travel increased to provide just the conditions needed to spark off the first epidemics. For some, the Flood is a parable that tells believers that the Lord punishes sinners and makes a covenant with those who follow in his ways. I concentrate instead on the geological evidence of past disaster and on the coming deluge as global warming gallops out of control.

The noble prose of the King James Version can provoke an emotional response even in a determined unbeliever (I am less confident about the New English Version) but as an explanation of physical events the Bible itself largely – and unsurprisingly given its age – fails. Science's job is to dispel mysteries, not to invent them, and, as I hope to show in these pages, often it does the job well, to give us an unexpected insight into the world and into ourselves.

My book avoids anything not open to rational explanation, whatever its importance to believers: the afterlife, the Resurrection, even God himself, for science can neither confirm nor deny notions based on spirituality alone. One correspondent bemoaned the fact that 'Jesus is not even in the index' (which would make an excellent subtitle in its own right). I do not consider the Good Book's myths and parables and I ignore the sense of mystery and the inner ecstasies claimed by many among the faithful, for such things are beyond the remit of what I know.

No thinking Christian today (and there are many) defends the notion that the Earth is six thousand years old, or that women descend from a man's rib-cage. To criticise such claims is to kick straw men when they are down and I do not even try. Fundamentalists make too easy a target – were there really dinosaurs on the Ark? – but some believers appear to be so annoyed by their statements that they throw out the blessed baby with the holy bathwater by denying that the Good Book has any contact at all with the real world. If those acolytes of the Church of the Holy Metaphor so despise the literalist view of scripture that they are unwilling to accept the factual accuracy of at least some of its verses, what *do* they believe? Not, it seems, very much. In their capacious philosophy none, or almost none, of the events recorded in the Testaments has a connection with the physical universe: instead, from Genesis to Revelation, they are to be read as symbol, allegory and myth.

I am the last person to comment on whether *The Serpent's Promise* is, as literature or as science, any good or not: but whatever its merits or lack thereof, the response from some Christians has led me to the unexpected conclusion that, sceptic as I may be, I have more faith in the Bible than they do. On the other hand, their system of thought is so different from my own that they may be right in the claim that they believe it from cover to cover, but in their own special way. Such issues are for theology, a topic about which I am ignorant and one which is absent from these pages. *The Serpent's Promise* is about science, not religion, and should be read with that limitation in mind.

INTRODUCTION

William Blake, *The Temptation and the Fall of Eve*

It is a foolish thing to make a long prologue, and to be short in the story itself.

II MACCABEES 2:32

When, long ago, I was a student in the Zoology Department of the University of Edinburgh, there stood in a niche on the main staircase a bronze statue of a chimpanzee. The object is still there. The animal wears a perplexed expression as it gazes at a human skull held in its fist. It sits upon a pile of books, one of which bears the name 'Darwin' upon the spine. On

1

the open page of another volume are incised the words '*Eritis sicut Deus*'. The phrase is a quotation from the third chapter of Genesis, as translated by St Jerome in the fourth century. As the Serpent persuades Eve to pluck the forbidden fruit the creature says: '*Eritis sicut Deus, scientes bonum et malum,*' which the King James Version renders as: 'Ye shall be as gods, knowing good and evil.'

Scientists are no more qualified than anyone else to comment on those two abstractions, but they have gained insights into the physical world rather more dependable than those of the Scriptures. Science (unlike the Serpent) has, in its brief history, lived up to most of its promises. It allows us to answer many of the questions that so mystified that Scottish primate and, as an incidental, gives this book a title.

The double helix and the mushroom cloud have joined the Cross, the Crescent and the Star of David as global icons. Like the ancient scribes, the people who invented those two images seldom ask new questions, but – unlike them – they do quite often come up with new answers. The topics studied by today's physicists, astronomers and biologists have obsessed mankind since long before their subjects began. God himself set problems about how the world works, as in his address to Job: 'Where is the way light dwelleth?', 'Hath the rain a father?', 'Where wast thou when I laid the foundations of the earth? … Whereupon are the foundations thereof fastened?' The Book of Proverbs does the same: 'Who hath gathered the wind in his fists; who hath bound the waters in a garment?' The response to such queries was, needless to say, that the universe had been called into being by the Lord himself and that its beauties were evidence of his existence, proof that 'The heavens declare the glory of God; and the firmament sheweth his handiwork'.

That logic is empty, but the questions posed to the unfortunate Job have become the raw material of research. Those who study Nature's ways today are interested, as were the sages of old, in the origins of the universe, of our planet, of living matter, of species and of mankind; and in the biology of sex and age and the possibility of eternal life – real rather than metaphorical – as against fiery doom in a decaying solar system. The repeated appearance of such themes in sacred texts, the Bible among them, is a reminder that each was a handbook to help comprehend the world and that each in its own way, and in its own day, succeeded.

The Good Book is many other things: a set of laws, some serious and some trivial, a history both real and imagined, a collection of precepts and of poetry, and an extended speculation about the glorious future that awaits those that accept its message. It sits firmly in the genealogy of ideas. Science is its direct descendant and the factual, if not the spiritual, questions asked long ago can be explored with the latest technology. This volume is an attempt to do just that, to scrutinise the biblical pages from the point of view of a scientist. In an attenuated version of its original, it tries to imitate the Testaments by weaving what might seem a series of unrelated facts into a coherent whole.

Religion itself can also be studied by members of my profession, on several levels; in terms of curiosity about this world and the next, and of the universal concern about the welfare of family, of nation, or of life as a whole. Research on the brain adds to the story, as do individual differences in genes and personality and in the social and intellectual background in which they are placed.

Powerful as the tools of science have proved, plenty of people dispute its findings on grounds of belief while others reject claims based on faith because they deny the truth, or are impossible to test. Even so, the attitude of the globe's billion or so agnostics and atheists to the doctrines of the devout majority has much in common with the views of the pious to the obdurate universe of fact, for each contemplates the other with a mixture of fascination and distaste. The idea that simple conviction can illuminate the physical world is devoid of interest to biologists, geologists and the like. Many who cleave to dogma have an equally negative attitude towards science, for they reject what they see as its claim to be a complete explanation of what surrounds them. As a result, many scientists have a furtive interest in what the fundamentalists are up to, and biblical literalists are often beguiled by science, if only to denounce it.

The twenty-first century has reawakened the serpent of superstition. Many have tried to strangle it while others prefer to stir the creature up. Polemical works for and against the power of belief stream from the press. Some attack its fundamentals while others do the opposite. Their attempts to do so have generated more than a thousand courses in science and religion in American universities (with a few scattered across the arid wastes of British academe).

This book is unlikely to appear on their reading lists, for most of their curriculum is beyond the capability, or logic, of science itself. In a covert attempt to accept that failing, some try to have a foot in both camps. They suggest that objective analysis can only go so far and that there must be another truth beyond. Alfred Russel Wallace – co-discoverer of natural selection – was certain that *Homo sapiens* had 'something which he has not derived from his animal progenitors – a spiritual essence or nature ... [that] can only find an explanation in the unseen universe of Spirit'.

Charles Darwin was dubious about such careless use of his ideas, but in a response to an attack on his colleague's claim, he noted that such statements were 'not worse than the prevailing superstitions of the country' (by which he meant Christianity). There he was right, but more than a century on, many still hold to Wallace's belief, as restated by Martin Luther King, that 'Science investigates; religion interprets ... the two are not rivals.' The notion that science and doctrine occupy separate, or even complementary, universes and that each provides an equally valid insight into the world seems to me unconvincing and is pursued no further here.

Even so, those involved in science can examine many of the claims made in the Bible in an objective way. *The Serpent's Promise* is not intended as a statement for or against the joy of sects; as an attack on, or defence of, Christianity or any other creed. My own views on the sublime, such as they are, play almost no part. Instead I attempt to stand back and take a fresh look at the sacred writings in a volume that tries to interpret some of its themes in today's language. The King James Version is more than six times longer than this work, and I have been obliged to omit many sections, such as the endless accounts of family trees and tribal battles, and the detailed instructions as to how to deck the Tabernacle.

This book begins, in the tradition of its model, with an account of the covenant between God and Man that began in Eden, with an attempt to trace the global pedigree from the inhabitants of that fabled land and from their real equivalents as revealed by modern biology. Genesis explains how the universe came into being, and I too gallop through history from the Big Bang to modern humankind. Eve's acceptance of the Serpent's promise led to original sin – to inborn imperfection – and biology has given us the ability to identify many of our own strengths and weaknesses even before birth (although the decision about what to do with such information has scarcely moved on since biblical times).

Her great error forced sin and sex to become close companions. That mode of reproduction ensures that life persists whatever the fate of those who transmit it. It means that sex levies a penalty in the currency of age and death. Decay, as a result, strikes us all, and long before it smote the patriarchs.

Soon after their demise, divine irritation with Man's degenerate ways led to the Great Flood, an event that has been tracked down in history even as today's fecklessness threatens us with a successor. The descendants of its few survivors fought among themselves even as they multiplied in number. In time, one chosen group, the Children of Israel, were taken into bondage, but in the Exodus found their way to a promised land, only to flee again after a political upheaval; an experience that the Jews have undergone again and again. DNA shows that every nation has the same history of exile and danger as humankind multiplied in number and filled the world. As it did, the Bronze Age Levant boomed and the earliest cities, Babylon included, appeared. They were accompanied by the first epidemics, and in the Book of Leviticus the priesthood set out rules of purity in an attempt to cope with a problem that remains a threat today. They also turned a jaundiced eye on which foods are wholesome and which are not as a reminder that diet is a potent statement of cultural and religious identity (and, as we now know, of health).

Visionaries, from Isaiah to Ezekiel, play a large part in the scriptural narrative and we now understand how some of their experiences arise, how the brain can deceive its owner and – perhaps – what lay behind some of the supposedly supernatural experiences of the biblical prophets and their successors.

The New Testament marks a great shift from the Old for it brings the scriptural narrative much closer to the modern world. Instead of a narrow focus on the doings of a chosen people and their implacable god, the Gospels emphasise altruism and inclusiveness, and the rewards to be gained in heaven by making sacrifices here on Earth in response to Christ's promise of eternal life. To believers, that philosophy explains the origin of devotion, and even of society (sceptics, in contrast, see religion as a confidence trick to concentrate power in the hands of a few). *The Serpent's Promise* ends with an account of today's attempts at a science of faith; and makes the modest proposal that now may be the time for the natural to supplant the supernatural as Man begins to make sense of the universe he inhabits.

About the supernatural itself, neither science nor this book can say anything. As the mathematician Laplace is said to have responded to Napoleon when the Emperor asked him why there was no mention of the deity in his volume on celestial mechanics: 'I have no need for that hypothesis.' An appeal to a supreme power added nothing to his understanding.

In spite of that succinct advice, Christians often try to accommodate the latest advances into their creed. From the heliocentric universe to the theory of evolution, new discoveries are woven into the texture of belief and used to bolster religion itself (the Big Bang, for example, must have been sparked off by God). Theological arguments of this kind depend on the idea that the existence of a final cause behind the universe can never be rejected. In the end – as Laplace pointed out – untestable mysteries of that kind are of no interest except to those determined to believe them.

The Frenchman's logic makes sense to his intellectual descendants, but would have seemed strange indeed to his predecessors, many of whom saw their task as no more than a step towards a comprehension of divine intent. The Bible plays as a result a large (and often neglected) part in the history of science, for many of its great figures were believers in a sense that modern minds find hard to understand. Isaac Newton was less interested in the 'Book of God's Works' – physics and mathematics – than in the 'Book of God's Words', the Bible. He wrote far more about philosophy than about physics, with a 300,000-word exegesis on the Book of Revelation which tried to prove that the Pope was the Antichrist (and on the way came up with the oddly comforting fact that 'We have no reason to suppose more

Apocalyptic Whores than one'). The rules of the universe had been set at least in part by an external agent: 'So then gravity may put the planets into motion but without the divine power it could never put them into such a Circulating motion as they have about the Sun.'

In the same way, Robert Boyle, father of chemistry, felt that the human body lives on even after death; 'its atoms are preserv'd in all their Digestions and kept capable of being reunited' (which explained the Resurrection). Robert Hooke, discoverer of the cell, saw the microscope as an attempt to restore the perfection of Man's senses, lost at the Fall, while Joseph Priestley, of oxygen fame, was equally sure that his *History of the Corruptions of Christianity* was worth far more than his research on gases. He hailed the French Revolution as a harbinger of the Second Coming and was forced to flee the mob as a result.

As even Isaac Newton might now be forced to concede, since his day the book of divine works, and that of words, have diverged in an essential way, for the first has moved on while the second has stayed more or less where it was. Reality is a stubborn thing and those who devote their lives to it are often forced to change their minds as the evidence changes. Religion, in contrast, depends on revealed and permanent truths. It evolves only in response to philosophical speculation and social pressure rather than as the result of new discoveries about divine intent. Three centuries after Newton, his most direct descendant, Albert Einstein, saw the Bible as no more than 'a collection of honourable but still primitive legends which are nevertheless pretty childish'.

The notion that physical or chemical laws could confirm its claims, or that they have themselves been put in place by a divine force, is alien to most of those who study them. The rise of scepticism in the United Kingdom (with, according to the 2011 census, a quarter of the population not defining themselves as religious and no more than one in five ever going to church except for weddings and the like) makes it hard to compare the attitudes of believers and their opposites. In the United States, where two thirds of the population trust in God with absolute certainty, half are sure that Jesus will soon return, and a majority say that they would be happier to vote for a Mormon, a Jew, or a homosexual as President than an atheist, the contrast is stark. In a survey of almost a thousand of the nation's top researchers, just two felt that the Bible should be interpreted as a literal truth, compared to a third of their fellow citizens. Two thirds of the American public, in contrast, say that they would continue to hold to a claim made by their Church leaders even if scientists were to disprove it.

In spite of their confidence in its message, their instruction manual has a chequered past. It is seen by many in the tradition of David Hume as 'a book, presented to us by a barbarous and ignorant people, written in an age when they were still more barbarous, and in all probability long after the facts which it relates, corroborated by no concurring testimony, and resembling those fabulous accounts, which every nation gives of its origin'. The first five sections – Genesis, Exodus, Leviticus, Numbers and Deuteronomy – comprise the Torah, which was, some believe, transcribed by Moses from the word of God. In fact, like its companions, it has multiple sources and was composed over many years. Its chapters are a palimpsest of manuscripts written and edited by known and unknown hands. Parts of the text emerged long after the actions it purports to describe. Some books are in most versions, while others have been excluded. Its tales are often inconsistent (as when, in Genesis, Man is created both before and after the animals). Some are supported by other evidence, while for many the ancient scrolls are the sole indication that the events recorded did happen.

Parts of the biblical message are deeply conservative, others radical. For more than a thousand years the only Christian text available was in Latin, and as most people could not speak that its mysteries were hidden from those told to believe them. The earliest complete English translation was made in 1382. It was suppressed. The first printed copy was that of William Tyndale in 1537. He was martyred for his pains.

A century later James I saw that an authorised edition was preferable to the demotic versions by then in circulation (some of which contained difficult terms such as 'tyrant') and sponsored a new translation. The King James Version of 1611 is written in Late Renaissance Newspeak, a language noble but purged of unacceptable ideas. It contained helpful phrases such as 'The powers that be are ordained by God' (which confirms that even the worst monarchs have a divine right to rule).

The King James has sold more copies than any other work in English, and in its language and its insight into an ancient and unfamiliar way of life has had a pervasive influence on

Western civilisation. Literature owes it an enormous debt: in Coleridge's phrase, 'intense study of the Bible will keep any writer from being vulgar in point of style'. He was right, and more recent translations, such as the New International Version used in many churches, are by comparison feeble. Their leaden lines dispel much of the mystery that still surrounds the King James and make many of its claims even less convincing than they appear when told in the language of four centuries ago.

Its contents have also had large effects, benign or otherwise, on politics and history. George Washington felt that 'It is impossible rightly to govern the world without God and the Bible' and his successor John Adams imagined that if 'a nation in some distant Region should take the Bible for their only law Book ... Every member would be obliged in conscience, to temperance, frugality, and industry; to justice, kindness, and charity towards his fellow men ... What a Eutopia, what a Paradise would this region be.' Some years later, George W. Bush announced that 'I feel like God wants me to run for President.'

The Serpent's Promise is by no means the first attempt to revise that great work. Noah Webster, of American Dictionary fame, was shocked by the lewdness of earlier versions and in his edition men have no stones and women no teats. Fornication has gone, as have legs (which are replaced by limbs). Thomas Jefferson went further, for in his adaptation he considered the miracle stories to be 'a ground work of vulgar ignorance ... superstitions, factitious, fabrications'. He cut out many of the wonders and dubious additions (and by that he meant the Trinity and the question of Jesus' divinity) in a search for its essence. The forty-six pages that remained, he was certain, 'extracted the diamonds from the dunghill' to give 'the most sublime and benevolent code of morals which has ever been offered to man'. His successors in the American Conservapedia movement are, in the spirit of the English monarch, even now at work on a similar project, with an attack on the liberal bias they see in modern translations ('volunteer' replaces 'comrade' throughout).

Many people have also pointed at what they see as objective truths within its pages. Job said that God 'hangeth the Earth upon nothing' (which might be a statement of a globe suspended in space) and spoke of the weight of the wind long before the notion that the air had mass, while Jonah fell to mountains at the bottom of the sea when cast overboard, as proof that he was the discoverer of seamounts.

Others interpret its tales in modern terms. The Plagues of Egypt happened around 1500 BC, as the climate warmed. They might have coincided with natural events: perhaps the stagnant Nile suffered an attack of red algae, which forced frogs onto land. Their corpses fed flies and mosquitoes, whose depredations caused animal disease and human boils. Then the Mediterranean island of Thera exploded, with a hail of fire. The rains that followed caused an outbreak of locusts, which blocked out the sun until, in the final disaster, the firstborn died because they ate mouldy and poisonous damp grain. Such claims are plausible and some may even be accurate, but there exists little direct evidence for any of them. Rather than attempting to explain particular scriptural events in detail I try to sketch out the Bible's larger themes.

My own attempt to reconsider the work is quite free of any taint of originality. I often warn students of the dangers of plagiarism but am myself a serial offender. I have in the past attempted to rewrite, or at least update, the entire oeuvre of Charles Darwin, and to use modern biology to test his ideas (which survive the process remarkably well). This attempt to do the same with the Good Book follows its structure in a less slavish fashion.

The Serpent's Promise is a book about dry fact, not theology (nor, God preserve us, philosophy). Its original has much more interest in the universe of the spirit than in the banalities of the physical world; a truth celebrated by many of its devotees. St Augustine criticised Godless curiosity about the tangible universe as 'the lust of the eyes ... a vain inquisitiveness dignified with the title of knowledge ... To satisfy this diseased craving ... people study the operations of Nature which lie beyond our grasp, when there is no advantage in knowing and the investigators simply desire knowledge for its own sake.' Science is that '*concupiscentia oculorum*'. Unlike its alternatives, it answers questions rather than just asking them. Progress depends on the hope that a theory may be disproved rather than on the acceptance of stated truth. Its enquiries know no limits, none of its explanations is complete, and authority, divine or otherwise, is never enough. Sometimes, as in the downfall of Newton's ideas as the foundation of physics from Higgs boson to cosmos, a whole subject collapses in the face of new evidence, but those whose temple has been thrown down do not wring their hands over the ruins, but dust themselves off and build a new one.

Its practitioners' most important principle is that they do not know what they will find. Without that admission, their subject could not exist, and humankind would be mired in ancient ignorance. The French physiologist Claude Bernard wrote in his 1865 book *Introduction to the Study of Experimental Medicine* that 'It is better to know nothing than to keep in mind fixed ideas based on theories whose confirmation we constantly seek.' He was right (although it took many years to persuade doctors to follow his advice).

The danger of doctrine is that its adherents seek confirmation of what they know; that their redeemer liveth, that there is no God but God and Muhammad is his prophet, or that one of a variety of other convictions must be true. A little knowledge is a dangerous thing, but certainty is worse. To the scientific method, faith is a vice; to believers, a virtue. In the Epistle to the Hebrews it is defined as 'the substance of things hoped for, the evidence of things not seen'. Such a notion is anathema to science, but central to religion, whose debates are held under the rules of the courtroom rather than the laboratory. A defence lawyer quotes only the evidence that favours his client and rubbishes the opposing case, however strong he can see it to be. A scientist may cling to his favourite theory for perhaps too long, but in the end must accept that he is wrong if the evidence goes against him. Doubting Thomas, who refused to believe in the Resurrection unless he was allowed to insert his hand into Jesus' wounds, was rebuked by the Saviour: 'Thomas, because thou hast seen Me, thou hast believed: blessed are they that have not seen, and yet have believed.' The dubious Thomas would make an excellent patron saint for scientists, but unfortunately the post has already been filled by St Albert the Great (who, needless to say, saw in the natural world a proof of the existence of God).

In biblical times people asked sensible questions about life, or geology, or the sky at night, and came up with what seemed sensible answers. Almost all were wrong, but their philosophy meant that there was no reason to revise them. Piety is – with its promise of life eternal – for optimists, while science is the home of pessimists, who search for ugly facts with which to destroy their (or at least their rivals') beautiful theories. They have brought doubt to the world and the world has gained as a result.

Not everyone agrees. Whatever the triumphs of modern research, a good portion of humankind still rejects its tenets because they conflict with their own opinions. Instead they prefer Martin Luther's assertion that 'All the articles of our Christian belief are, when considered rationally, impossible and mendacious and preposterous. Faith however, is completely abreast of the situation. It grips reason by the throat and strangles the beast'; a statement which has at least the virtue of honesty.

That way of thinking has given rise to what some see as a new Age of Endarkenment. The word contrasts the seventeenth-century outburst of intellectual creativity with that manifest in today's faith healing, Jesus' face in tomatoes or on toast, and the rest of the medieval clutter which dominates so many lives. Its adherents insist that no attempt to understand the universe that omits the spiritual can be complete, whatever the advances of biology, chemistry, physics and the rest, which may be why 40 per cent of the United States population – plenty among them students of biology or medicine – deny the truth of evolution (and why more than twice that proportion of Pakistanis, Egyptians and Malaysians agrees).

Millions more reject the notion of man-made climate change because they do not like the idea. I find such views impossible to understand. Why listen to a perjurer paid by the oil industry when discussing global warming, or train to become a biologist and at the same time deny the very foundations of the subject? To do so is like doing a degree in English while rejecting the existence of grammar, or in physics with a rooted objection to gravity: it makes no sense.

I sometimes wonder whether those who pour their inane doctrines into their pupils' ears ever consider the damage they do; not to my profession, but to theirs. Why, when a student begins to learn the simple and credible facts, rather than fantasies, about how life emerged or the atmosphere works, should he swallow anything else that his pastor, his rabbi or his imam has told him? Why build a philosophy based on fixed untruths, when we have so many truths, and so much still to find out? There, science cannot help.

The closer scientists draw to the spiritual the less precise their statements become, but I hope to make the case that reason is a better way to understand the physical universe than is faith; that whatever the historical importance of the latter, or the solace it offers to some, that science is a more consistent, universal and satisfying tool with which to organise human lives. A few may be converted from one view to another while more will see no reason to change their opinions and some may be no more than irritated by my presumption. Whatever their response, I hope that readers will learn something from this endeavour to put unfamiliar facts into familiar context. The illustrations at the chapter heads are by William Blake, who demonstrates, better than almost anyone else, the power of sacred imagery to move even those who do not share his convictions. Much of his oeuvre itself is based on a radical new interpretation of the Good Book. Blake expels the corrupted God and replaces him with his divine son, and does so with such genius that I forgive him his statement that 'Art is the tree of life: Science is the tree of death.'

My own attempt to emulate his work, feeble as it may be in comparison, flies in the face of scriptural advice; as the Book of Revelation puts it, 'For I testify unto every man that heareth the words of the prophecy of this book, If any man shall add unto these things, God shall add unto him the plagues that are written in this book: And if any man shall take away from the words of the book of this prophecy, God shall take away his part out of the book of life.' *The Serpent's Promise* takes that risk and I await the consequences with interest.

PROLOGUE

THE DESCENT FROM ADAM

William Blake, *The Primaeval Giants Sunk in the Soil*

There were giants in the earth in those days; and also after
that, when the sons of God came in unto the daughters of
men, and they bare children to them, the same became
mighty men which were of old, men of renown.

GENESIS 6:4

Genesis was the world's first biology textbook. Its obsession
with ancestry is reflected in much of the rest of the Bible. The
word 'begat' appears a hundred and forty times in its pages,
while 'son' gets two thousand mentions in the Old Testament

alone. Judaism, the most genetical of all creeds, is built on the assumption of shared descent from Abraham, who had himself a direct tie with Noah's son Shem and was as a result a successor to Adam.

Christianity, too, links God's immortal son to Abraham in a statement of continuity. Matthew's Gospel describes itself as 'The book of the generation of Jesus Christ, the son of David, the son of Abraham' and rehearses the three sets of fourteen ancestors that separate Abraham from King David, David from the exile to Babylon, and from then to the Saviour. In the Middle Ages, the image of the Tree of Jesse was popular. It traces Christ's descent back to Jesse of Bethlehem, the father of David, and is based on the prophecy in Isaiah that '... there shall come forth a rod out of the stem of Jesse, and a Branch shall grow out of his roots: And the spirit of the LORD shall rest upon him ...' (part of that notion comes from the similarity of the Latin word for rod, *virga*, and that for the Virgin, *Virgo*). A modern work of perhaps lesser merit also points at the unshakeable fascination with shared blood, for *The Da Vinci Code* describes the marriage of Jesus to Mary Magdalene to provide a direct connection between a modern Frenchwoman and the son of God. All this is part of a deeper message; of the need to establish a covenant between sinful Man and the purity of Eden before the Fall, and – for Christians at least – to enter an afterlife that recreates those days of innocence.

Humankind's moral decay since Eve accepted the fateful fruit was, many once thought, accompanied by physical decline. The Koran sees Adam as a sixty-cubit monster (a cubit is the length of a forearm) with, to match the world's spiritual degeneration, much shrinkage since his day. That notion was supported by the bones of mammoths that hung in some churches and allowed experts to estimate Adam's stature as twice that put forth in the Koran. There were, as the Good Book itself says, once giants in the Earth; and as so often it is right. Those titanic figures illustrate, better than almost anything else, the power of science to illuminate myth.

Until not long ago such people were creatures of wonder and of veneration. They crop up in the Bible Lands as a race or tribe ('Giants dwelt therein in old time; and the Ammonites call them Zamzummims'), as rulers ('For only Og king of Bashan remained of the remnant of giants; behold his bedstead was a bedstead of iron . . . nine cubits was the length thereof, and four cubits the breadth of it, after the cubit of a man') and as powerful enemies ('And there went out a champion out of the camp of the Philistines, named Goliath, of Gath, whose height was six cubits and a span').

Spies sent into Canaan reported that for the biblical giants '. . . we were in our own sight as grasshoppers, and so we were in their sight.' Some claimed that such creatures were the descendants of rebellious angels who had mated with women; a bold notion but one rejected by those who accept Jesus' statement that angels do not marry. Others thought that the monsters were the spawn of Seth, third son of Adam and Eve, conceived to replace the murdered Abel. Their ancestors had polluted their heritage by marrying into the lineage of the killer Cain, and were punished with the repeated appearance of freaks of nature among their descendants. The giants were swept away in the Great Flood but make a brief and unexplained return later.

The Good Book's fascination with ancestry finds a match in the obsession still felt by many people about their own pedigrees. For them, as for the Israelites, shared descent acts as a badge of membership of a family or a nation, and even as a mark of adherence to a particular faith.

Genetics is poised to answer many of their questions. It can now read off the three thousand million pieces of information coded into the four chemical units or 'bases' of the double helix of DNA, the letters A, G, C and T, in a few hours. By so doing it can chase down the ancestry of anybody, tall or short. We are creatures of great diversity. A new survey of the complete sequence of the protein-coding genes (themselves only a small part of the DNA) from a thousand people from across the world reveals forty million differences in the individual letters of the genetic alphabet and over a million separate insertions and deletions of sections of double helix. The constant rearrangement of these changes through sex, and the endless input of new errors by mutation, means that everyone alive today is different from everyone else, and from everyone who has ever lived, or ever will live. More remarkable, every sperm and every egg made by every man and woman is different from all others, as has been each of those fecund cells made in their infinite billions since Adam met Eve. The double helix, the universal record of the past, can as a result track down the pedigrees of the inhabitants of Eden, of the biblical giants, and of everyone else.

The giants are still around. The world's tallest man, *The*

Guinness Book of World Records tells us, towers two and a half feet above my own modest frame for he stands at eight feet three inches. The young Sultan Kösen also boasts the globe's biggest feet at fourteen and a half inches long. His plight has been explored not with legend but with biochemistry.

Sultan Kösen is Turkish. A group of Istanbul businessmen once explained to me that the idea of human evolution was of course correct – except for Turks, who had emerged from titanic figures who once roamed Anatolia. That conversation took place some years before their fellow-citizen claimed his title but they would no doubt have welcomed his success as proof of their theory.

The people of Turkey are not alone in their claim of an elevated past. Finn MacCool was the gargantuan leader of the armed band that guarded the High Kings of Ireland. He was the architect of the Giant's Causeway, the rocky point that stretches from Ulster towards Scotland. It appeared when he built a path to allow him to fight a rival across the water (his opponent used it to arrive in Finn's native province while the Ulsterman was asleep, but was warned by Finn's wife not to wake what she called her 'baby', whereupon the Scot fled for fear of the supposed infant's father).

Finn MacCool may have had a real existence. Tales of tall men appear again and again in his native land, often enough to suggest that they are more than fantasy. Their remains have surfaced on several occasions and such people still thrive on Irish soil.

In 1895 the *Strand* magazine reported the excavation of a twelve-foot skeleton in County Antrim. The bones were lost in transit to the London and North-Western Railway Company's depot and have not been seen since. Fortunately, other evidence has survived.

In the eighteenth century extraordinarily tall men were popular items of curiosity (as they still are). One such, Charles Byrne (his surname the anglicised form of O'Brien), was born in County Tyrone in 1761. He claimed to be eight feet two although his real height as measured from his skeleton was six inches less. At the age of nineteen, he put himself on show in London and made a career from those happy to pay to see a freak of Nature. A contemporary print shows him in company with two equally tall twins who lived near his home and claimed to be his relatives. Charles Byrne was also kin to another lofty Irishman, Patrick Cotter O'Brien of Cork. He too exhibited himself on this side of the Irish Sea, declaring himself to be a descendant of the colossal King Brian Boru. Accounts of such people were so widespread that James Prichard in his 1813 book *Researches Into the Physical History of Mankind*, the work that founded British anthropology, wrote that 'In Ireland men of uncommon stature are often seen, and even a gigantic form and stature occurs there much more frequently than in this island: yet all the British isles derived their stock of inhabitants from the same sources. We can hardly avoid the conclusion that there must be some peculiarity in Ireland which gives rise to these phenomena.'

What that peculiarity might be he had no idea, but now we know. The Irish Giant took to drink and died at twenty-two. He had a great fear that he might be dissected and demanded that he be buried at sea, but his bones were too valuable for that; as a journal of the time put it: 'The whole tribe of surgeons put in a claim for the poor departed Irishman and surrounded his house just as harpooners would an enormous whale.' The winner was the anatomist John Hunter, who paid the huge sum of five hundred pounds for the corpse. He filled Byrne's coffin with rocks to fool those who hoped to honour the unfortunate man's final wish. The skeleton is now on display in the Royal College of Surgeons, where it is likely to remain in spite of calls for its return to the land of his fathers, or even for it to be condemned to the deeps.

In 1909 the bones were examined by the American neurosurgeon Harvey Cushing. He noted that a section of skull at the base of the brain was enlarged and suggested that the Irishman may have had a tumour on his pituitary gland. People with that condition, we now know, may make too much growth hormone. Should the problem be delayed until after puberty the bones of the skull and elsewhere become heavy and thickened. That leads to a disease called acromegaly. It changes the patient's appearance and has unpleasant side effects such as diabetes, arthritis and more. If it begins in childhood the main effect is on height, for arms and legs grow far faster than they ought and the children mature as modern-day versions of Charles Byrne. Sultan Kösen himself has the disorder but with the appropriate drugs has managed to put a stop to his growth.

A certain form of pituitary tumour has a strong inherited component. The gene involved has been tracked down and the altered form found within four Irish families, each of which has a history of acromegaly or of extreme height. Every one of the fourteen cases shares just the same change in the DNA, together with a long stretch of almost identical sequence on either side, as a hint that it descends from a common ancestor. Nucleic acid extracted from the teeth of Charles Byrne bears the identical anomaly, as proof that he too (and, no doubt, his giant contemporaries) belongs on the pedigree. Several hundred of his extant relatives have inherited the mutation (although, for reasons unknown, many have normal patterns of development). An Indian giant has an identical mistake in the gene itself, but carries a different arrangement of DNA letters around it, as proof that the damaged gene has arisen on separate occasions in different places. No skeletons of Og King of Bashan or of Goliath have been preserved, but the bones of the Egyptian Hen-Nekht, a nobleman of around 2500 BC, a thousand years before Og's day, show that he too was eight inches taller than average. His skull had signs of the Irish disorder ('extraordinarily massive, remarkably long, and with marked grooves and ridges'). The biblical giants may hence have been related to each other, to an Egyptian who lived fifty generations earlier and, perhaps, even to the gigantic Adam via his son Seth and his tribe of monsters.

Educated guesses (and they are no more than that) about the rate at which the few changes in the double helix on either side of the site of the Irish mutation build up hint that their common ancestor lived around fifteen hundred years ago. That is the time of the High Kings, when giants such as Finn MacCool begin to play a part in the nation's history. Half a millennium later, the tenth-century Goliath Brian Boru brought their dynasty to an end.

The ability to connect an eighteenth-century Irishman to his modern descendants shows how the double helix can link the past to the present. The Bible uses written records to do the same. It traces everyone's pedigree from Adam and Eve, from the heroes of the Good Book, and from a great host of other people, to link the Children of Abraham into a web of shared kinship.

To search out descent has become a renewed obsession among amateur genealogists, many of whom appeal to websites with names such as 'Genes Reunited' or 'Ancestry.com' in the search for the past. Like most geneticists I have little interest in my own ancestors, real or imagined, but for many people the hunt for shared blood is almost a mania. The double helix gives a heritage to those who have lost touch with their roots and the hunt for those who have gone before has become big business.

It began in the year 2000 with a company called Family Tree DNA. I was approached at that time by a private entrepreneur who wanted to know how he might make money out of the new fashion. I suggested that African-Americans, whose inheritance had been stolen from them in the days of slavery, might be keen to delve into the past. What, he asked, would be needed? The real demand was for a map of gene distribution in West Africa. As my colleagues had just started to work in that neck of the woods I accepted his offer to fund the project.

A few days later came a phone call. He had a question: was there any way that subscribers could check a statement that their families had come from, say, Cameroon rather than Nigeria? No, I said; and with joy in his voice he said he would not need our help for any company he might decide to establish could come up with a form of words that would satisfy its clients with no need to spend any money on fieldwork. I made my excuses and left; but since then at least forty companies have offered to do the job. They provide a scan of thousands of variable sites across the whole genome, sometimes for less than a hundred dollars and, for only a little more, promise to introduce you to dozens of relatives you never knew you had.

The Genesis account of descent begins with Adam and ends ten generations later with Ham, Shem and Japheth, the sons of Noah. The even longer trees in the First Book of Chronicles – upon which Abraham and King David, ruler of Jerusalem, both appear – are said to have provoked nine hundred camel loads of commentary. Millions of people claim a place upon them. Many Jews can follow their descent back to Rashi, an eleventh-century sage who established, to his own satisfaction, a direct tie to David. Rashi's modern representatives include Sigmund Freud, Karl Marx, Felix Mendelssohn, Yehudi Menuhin and all the Rothschilds.

Distinguished as they might be the scions of Rashi are arrivistes. The Chinese sage Confucius lived in the fifth century before Christ. He has a million and a half recorded living descendants, each separated from their ancestor by eighty or so generations. Confucius himself has a well-attested tie with the Emperor Tang of Shang, who was born in about 1675 BC, half a millennium before his Israelite equivalent, King David.

Few Britons can match that. Most of us can chase our heritage through the records for no more than a few score years. The Joneses in their multitudes are a dead end for ancestor-hunters but I did once pursue my mother's line to a William Morgan, born in 1759. He lived on the farm in West Wales still occupied by my relatives and was the great-great-grandfather of William Morgan, the nineteenth-century farmer-preacher to whom this book is dedicated.

Unremarkable as any clan – Jones or not – might be, each of its members has, within every cell of his or her body, a complete set of links to their ancestors and, in the end, to everyone who has ever lived. The record acts as a timer, for as the double helix is copied it is, like a scroll written out again and again by the scribes, corrupted by errors. Over the years, such mutations build up at a more or less steady rate. DNA can as a result date any pedigree on a scale from decades to millennia. The ability to read off the double helix at speed makes it possible to compare the sequence of a child with that of its parents to measure the mutation rate. It gives an average of a little more than one error in a hundred million per DNA letter per generation – which means around sixty mutations in a typical newborn.

The Good Book itself has suffered many such errors. Several editions have gaffes that identify their moment of publication. The Adulterous Bible of 1631 omits the word 'not' from the Seventh Commandment which then reads, with some optimism, 'Thou shalt commit adultery.' Less outrageous was the 1717 version that mutated the Parable of the Vineyard into the Parable of the Vinegar. The Sin-On Bible of a year earlier converts 'sin no more' to 'sin on more'. Such shifts represent in turn mutation by deletion, by changes in individual letters, and by inversion of their order, all of which have parallels in DNA. Other slip-ups were no doubt corrected by proof-readers before they became public.

Before the invention of movable type many more blunders were made as a new copy of a manuscript followed the last. More than twenty thousand handwritten versions of the New Testament are known, and no more than a few are the same. If enough versions of a sequentially corrupted manuscript have survived, a tree of errors can work out what the original might have looked like, and even estimate when it was written.

The Great Stemma is a parchment that shows the descent of Jesus and of hundreds of other figures, all traced from Adam. The document is mentioned in scrolls from the earliest days of Christianity, but the original is long lost. Around twenty versions have survived, the oldest from the tenth century and the youngest from three hundred years later. Even the earliest diverge from each other as proof of their descent from a common precursor while later copies become more and more corrupt. If the lost manuscripts had accumulated mistakes at the same rate, the original can be dated to the fourth century, when the Bible was translated into Latin. Mutations in the double helix can, in much the same way, place people into historical context.

The urge to do so goes back to ancient times. In the Levantine villages of the Bronze Age where Judaism finds its roots, membership was from the early days based on an assumption of kinship. One of the earliest mentions of the nation emphasises the importance of inheritance. An Egyptian stele commemorates a great victory in 1207 BC with the statement that 'Israel is laid waste; his seed is not.' Relatedness did not always draw people together, for Genesis tells how brothers – Cain and Abel, Esau and Jacob, together with various other relatives – murdered each other, cheated siblings out of their inheritances, sold them into slavery, or drove them into exile; but then things began to change.

The first Jewish state was founded more than three thousand years ago from scattered tribes whose later cycles of quarrels with God, delivery into the hands of enemies, and return to the fold, are described in the Book of Judges. In the early years household gods – the 'gods of the fathers' – competed with Yahweh – first mentioned in Moses' time – for the people's attention. In time their numbers reduced, and even Baal, with his Golden Calf, fell from favour. Soon, with the emergence of the Kings of Israel, Yahweh became the national god, although a few competitors lingered on.

Under their Kings the Jews united but, in the ninth century before the Common Era, split into two, Israel in the north and Judah, with its capital Jerusalem, in the south. Many verses of the Bible discuss their squabbles but in time they were reunited. For a period the nation's borders stretched far beyond those of today and under Solomon its citizens built the First Temple in Jerusalem. Even that statement is disputed by many, and attempts to excavate on the Temple Mount to find firm evidence have been banned by the Islamic authorities. Excavations on Mount Gerizim, now in the occupied West

Bank, suggest to some that the Samaritans, who see themselves as the true descendants of the founders, built the first great place of worship there instead. It was, they claim, demolished by Jewish enemies and the Samaritans more or less written out of history. The northern state had suffered its own calamity when the Assyrians destroyed its capital and took its people into exile. That event gave rise to another hereditarian legend, that of the Lost Tribes; families named after Simeon, Ephraim, Zebulun and the rest, the issue of Jacob.

Jerusalem was ruled for centuries by a clan of monarchs, the successors of David, the victor over Goliath. The devout saw their metropolis as the abode of God himself, a sacred capital where that dynasty would rule for ever (an idea shattered when the Temple was overthrown by the Babylonians six centuries before the birth of Christ).

Its citizens were among the many to assume that mankind was divided into distinct tribes and that within each, everyone had common descent. The notion of a shared and exclusive kinship became the rationale for unity and an excuse for war against those who did not belong. A fascination with bloodlines has had effects just as malevolent ever since.

Man is a classifying animal, of himself most of all. The great subdivisions of humankind were once said to be the African, European, Asian, Malayan and Native American races, with groups such as the Europeans further divided into Aryan, Slavs and Nordics, or Welsh, English and Scots (even a Cockney race was once proposed). Each section or subsection was a branch or a twig on the tree of relatedness. Implicit in such claims is the notion that in the grand sweep of history, peoples of unblemished heritage left their homelands and mixed to make the nations of today. Such mongrelisation was to blame for most of the problems of the world.

Modern Israel still defines itself, at least in part, by descent. Its obsession with biological continuity means that, for its most devout citizens, birth control is unacceptable. Some would even be happy to accept cloning if it were needed to allow their line to persist. Adoption is under strict control and is forbidden if the child comes from outside the nation's borders. Israel is the only country to offer almost unlimited fertility treatment, with a strong emphasis on in-vitro fertilisation (which preserves the genetic link between both parents and their child) as distinct from artificial insemination by donor, which works better but breaks the sacred chain of ancestry. IVF is offered to any woman under forty-five, married or not, until she has had two children with her partner. As a result, Israeli women are the world's greatest users of such technology.

A decade and a half ago I passed through Tel Aviv Airport after a trip to collect DNA from Palestinians. I was for some reason picked on by Security and grilled about every item in my luggage. At last they came to a box filled with plastic tubes and asked, with some suspicion, what it was. Irritated by twenty minutes spent in an attempt to justify my existence I said, with some asperity, 'Arab spit.' This gave them pause but after a brief explanation of what I was up to the atmosphere lightened, for the officials were happy to hear about the then little-publicised pastime of studying history with genes. They were particularly interested in whether the double helix might reveal a common heritage for Jews that would set them apart from the descendants of other tribes (or, at least, from the modern inhabitants of Palestine). A scientific proof that they were the true scions of the Kingdom of Israel would be welcome indeed. In my new and amicable relationship with the uniformed branch, I did hint that perhaps it was, for reasons from recent history, a mistake to define Jews as members of a distinct bloodline, but they did not seem much concerned.

Shared ancestry was used to justify the greatest crime of the twentieth century. The Roman historian Tacitus claimed two thousand years ago that the *Germani* were, unlike any other nation, still a pure group united by biology (in part, he thought, because nobody else in their right mind would want to migrate to a place with such a terrible climate). They were a noble people: 'No one in Germania laughs at vice ... and good habits are here more effectual than good laws elsewhere. The tribes of Germany are free from all taint of inter-marriages with foreign nations, and they appear as a distinct, unmixed race, like none but themselves. ... All have fierce blue eyes, copper-coloured hair and huge frames.'

His ideas stuck. 'Read Tacitus,' said the eighteenth-century philosopher Johann Gottfried von Herder. 'The tribes of Germany have not been dishonoured by intermixture with others, they are a true, unadulterated nationality which is the original of itself. Even the formation of their bodies is still the same in a large number of people.' In 1853 Count Joseph Arthur de Gobineau published his *Essay on the Inequality of the Human Races*, the work that introduced the term 'Aryan' into anthropology. That group represented the highest point of human excellence. It had spread to found the cultures of ancient Egypt, Rome, China, Peru and (needless to say) Germany. The Aryans (the 'Nobles' in the ancient Asian language Sanskrit) were members of a grand and extended pedigree. They came, he imagined, from a distant and romantic landscape and their blue-eyed descendants had colonised much of the globe. The idea that all advanced civilisations derive from the Aryans was picked up by Nietzsche, to whom the Nordics – the 'blond beasts' – were the purest specimens of all. The Germans were remnants of a great people who might one day expel, with what rigour was necessary, the lesser breeds that had infiltrated its homeland. The idea became notorious with the twentieth-century attempts to put that theory into practice.

DNA means that scientific fact has succeeded historical fiction. Now, any nation can use the double helix to chase down its past. The ancestry companies have gone so far in their expertise (and income) that the data they generate is often more precise than that produced by geneticists. They are much appealed to by those who hope to prove a personal tie to a noble person, or a noble people, from ancient times.

The biblical seekers for shared blood, like many of their successors, were interested in just a single track through history. It passes from fathers to sons. Certain genes, too, are transmitted through males alone, while others go through females. Each can be used to search out ancestors in a journey that leads, as it must, to Adam and to Eve, the progenitors of all men, and of all women, on Earth.

The male line is tracked through the Y chromosome. Many Chinese can recite their fathers' pedigree back through twenty or more steps. A DNA test of the chromosomes of two Cantonese who asserted descent from the same individual and were separated by thirteen generations showed that they were correct in their claim. So, no doubt, are millions more of their fellow-countrymen, those who declare a tie with Confucius included.

Some of the scions of the philosopher share another clue of common ancestry, for they bear the surname 'Kong'. Western names, too, connect the present to the past. The gigantic Charles Byrne's surname in its native form O'Brien linked him both to his enormous contemporary Patrick Cotter O'Brien and to the ancient colossus, Brian Boru. Several carriers of the pituitary mutation still bear that title.

Surnames and Y chromosomes both pass through history without sex. Each faces the same problem. If a man has no children at all, or no sons, his surname and his Y will both be lost (as genealogists say of the latter group, they have 'daughtered out'). As the years go by this process continues until at last just a solitary version of both name and gene is left. It can be seen at work in Britain. Some four hundred native names are each carried by more than ten thousand people, but there are hundreds of thousands of much rarer titles. Those with fewer than two hundred bearers – and an uncertain future – include the families Edevane, Ajax and Slora; while the Pauncefoots and the Footheads have disappeared within recent times.

Surnames collapse a whole pedigree into a single line of descent shared by all the sons, grandsons and later male descendants of the man who founded it. They have a chequered history. Japan did without them until the nineteenth century, when a great variety sprang into existence at the command of the Emperor Mutsuhito. Over a hundred thousand are still in widespread use – one for every thousand citizens – as few have yet been lost through death or daughters. Welsh surnames are no more than three hundred years old and their English equivalents around twice as ancient. Most Chinese versions can, in contrast, be traced back for five thousand years. China has the world's most frequent title, Li, and has fewer surnames, in relation to its size, than almost anywhere else. The figures are remarkable. A fifth of the country's population – around three hundred million people – share three names as evidence of how few of its male lineages have maintained themselves. In France the average number of bearers of a particular surname is seventeen, in Britain twenty-eight and in Ireland sixty-three. In China the number is seventy thousand.

The biblical scribes tried to include every significant male on the line through the past, and the more the better. Its verses are filled with statements such as 'The children of Shem; Elam, and Asshur, and Arphaxad, and Lud, and Aram. And the children of Aram; Uz, and Hul, and Gether, and Mash. And Arphaxad begat Salah; and Salah begat Eber. And unto Eber were born two sons' and so, endlessly, on. The Judaic habit of calling oneself after a father, grandfather and the rest, with the prefix 'ben' as a link between generations, resembles the Welsh 'ap', used in the same way. Today's Prices, Pughs and Proberts have no link apart from their descent from a Rhys, a Huw or a Robert whose children were obliged by the English to take up their father's identity. Problems with adoption, deliberate shifts of surname (often in order to receive an inheritance) and illegitimacy (even if its incidence has been, DNA tests show, less than 1 per cent in most European populations) also muddy the genetical waters.

The biological headquarters of male descent is a reduced and battered version of what was once the equivalent of an X chromosome. Although it contains sixty million DNA bases it has no more than around sixty functional genes (not least the tiny structure that dooms its bearer to be a man) but, with the exception of some strange stretches that are perfect mirror images of each other, is otherwise full of duplication and decay.

Over the generations the Y builds up errors, on two distinct scales. The chromosome contains lots of short repeated segments. Such structures are unstable and, as they are copied, their numbers go up or down. In the two Chinese men separated by thirteen generations there had been four mutations of this kind. Overall, the mutation rate for these repetitive sections is as much as one in a thousand per generation, which means that differences build up at some speed. Other variants involve much rarer shifts in the individual letters of the code. Mistakes of that kind happen just once or twice in human history and divide the world's Y chromosomes into large and stable groups. Within each, changes in the repeated segments mount up to give every local lineage an identity. The system is rather like surnames themselves. German titles are distinct from those in Spain, but within each country there is plenty of variation. The unstable sections can be used to search out genealogies over a few hundred years, but evolve so fast that the distant past is obscured by the fog of mutation. The rarer changes in the single letters of the chromosome tell the tale of more ancient kinship.

To mix metaphors, molecular surnames also have elements in common with telephone numbers. The more digits they contain, the better they track down their owners. In Alexander Bell's day, a local number had no more than two or three digits, enough to cover the few subscribers in a village or small town. Hundreds of scattered people in different places across the country would share the same number. In cities it was soon necessary to use more digits; in most places seven, the first three of which divided the place into smaller sections, while the remainder identified individuals within those areas. Again, there were repeats in different cities. Four additional figures then gave every British phone a unique personality. For global use even more have been added, to give, for most places, fourteen altogether, enough for several thousand telephones for everyone on Earth, which should do for now.

The first Y chromosome trees, a decade or so ago, were based on no more than half a dozen individual letters, but the most recent – many from genealogy companies – have fifteen times as many and a few have several hundred (and, quite soon, the complete sequence of individual chromosomes will be available). They can be used to place any man onto his local, national or global tree of descent.

An identity conferred by name and by chromosome often overlaps. Almost all the hundreds of Britons called 'Attenborough' have either just the same version of the Y or a form changed from it by a few minor shifts. Each must descend from the same individual, who may have lived several centuries ago near a fort or 'burgh' that was, given the present distribution of the title, somewhere in the English Midlands. Other fort-dwellers might have taken up the same surname, but their lineages died out when, in a certain generation, their male descendants either had no boys, or no children at all. All today's Attenboroughs descend from one or a few males. The Smiths, those innumerable workers in metal, and the Joneses, the sons of many Johns, are in contrast a mixed lot for their label appeared on dozens of occasions. Their Y chromosomes are just as bastardised.

On the other side of the Irish Sea, in contrast, those who bear certain abundant surnames, such as O'Brien (the sons of Brian), tend to share a Y. They are the descendants of noble families founded at a time when a few powerful males granted their favours to many females. One man in six in the northwest of the island bears more or less the same version of the chromosome, perhaps because they descend from the High Kings of Ireland, whose males spread their seed to a large constituency. Charles Byrne lived nearby and he too may have shared that identity.

Irish surnames have an unexpected resonance in English history. In the eighth century and later, Ireland suffered a series of invasions by the Vikings. In time they established settlements and mingled with the locals, which is why some of the kings had Nordic titles such as Magnus or Olaf. I lived as a schoolboy on the other side of the Irish Sea, on the Wirral Peninsula (which my friends and I used to refer to, with some bitterness, as Liverpool's Left Bank), a dull industrial suburb with almost no discernible personality. The Wirral has some odd place-names – Thingwall, Irby, Raby, Meols and more – which interested us not at all. They are in fact the relics of a forgotten Viking enclave, founded by warriors driven out of Ireland a thousand years ago. My local golf course was the site of the Battle of Brunanburh of 937, when the invaders were defeated and where England was born as a single country. The event finds a place in the Icelandic sagas and Alfred Lord Tennyson modernised the Anglo-Saxon version of the tale ('Bow'd the spoiler/Bent the Scotsman/Fell the ship-crews/doom'd to the death'). Most of the golfers do not seem to be particularly concerned.

A check of the Y chromosomes of men with surnames such as Irby and Raby (whose medieval families must have lived in the ancient Nordic villages) shows them to have strong ties with Scandinavia and hence to be descendants of Vikings. The double helix gives the Wirral Peninsula a romance that otherwise it lacks.

Useful as names are, some who believe that a shared surname is proof of common ancestry are mistaken. The Prophet Muhammad had no sons, but through his daughters he did have grandsons, Hasan ibn Ali and Husain ibn Ali. The Sayyid kindred is, many of its members suppose, descended from the Prophet down the male line from one of those two. Some once held high office and were even exempted from taxation. If they are correct they should, like the Attenboroughs, share a Y chromosome, but they do not. Sayyid status is based on tradition rather than on biology.

Other self-styled aristocrats have an identity just as ambiguous.

Iranians tell themselves, with some historical evidence, that they descend from an ancient people called the Arya. About a quarter of all Indians (in particular those of the upper castes, in part derived from invaders from what is now Iran) also claim that bond. Many members of those groups do indeed share a similar set of Y chromosomes.

The Germans too claimed kinship with a pure, blond and talented race from a distant mountain homeland. That argument was used as an excuse to destroy lesser breeds in an attempt to restore the Aryan nation. In 1941, the Germans invaded Russia to fulfil that dream but were – like so many before them – beaten back.

Nazi racial theory was as confused as its military strategy. Russia has three major Y lineages, each marked by a shift in the single-letter cues of identity, the most frequent borne by half of its men. That sequence is close to that of the historical Arya of Iran but in Germany is almost absent. Instead that nation has many copies of a different version of the Y, which is itself abundant in the Middle East. The ancient Aryan warriors hence have more in common with modern Slavs than with Germans, and the Teutonic purists who saw the Arya as their own ancestors had on average a closer tie with the Jewish men they despised.

The feminine path through the past can also lead to unexpected places. Lineages that pass through daughters once received less attention than did those of sons although that has changed (and even Confucius' pedigree has been revised to include – for the first time – women). Its signposts are the cellular structures called mitochondria. Those energy machines are passed through the egg to sons and daughters, but daughters alone hand them on. Each contains many copies of a small circle of DNA. Like the Y chromosome it builds up differences at some speed. As it does it tells the tale of mothers gone by.

Often, their histories diverge from that of their partners. The Parsis are Zoroastrians whose legends tell that they were expelled from what is now Iran in around the tenth century AD. Their Y chromosomes show that they do indeed have ties to the Arya of that region. Many of their female lineages, in contrast, find a closer match in Gujarat in North-West India, where Parsi men may have found mates among the locals.

Male invaders have often mated with the women of a conquered nation. At the southern tip of Africa, the 'Cape Coloureds' trace their roots to the seventeenth-century arrival of the Dutch East India Company, and to the later importation of slaves from elsewhere in Africa and from what is now Malaysia. The local Khoi-San tribes resisted the white colonists, who in revenge killed many of their men and drove others out. For more than a century, the Cape Colony saw an influx of European males, but of almost no females. Among the slaves, too, there were six times as many men as women. Khoi-San women faced pressure to enter into relationships with foreigners. Many did, but there was almost no sex between Khoi-San men and alien women.

Overall, about a third of today's Cape Coloured genes are Khoi-San, a third are from Black Africa, and the rest come from Europeans and Asians (which explains their wide range of skin colour). However, six out of ten mitochondria are of

Khoi-San origin, compared to just one in twenty that traces itself to a European. In contrast, just a twentieth of their Y chromosomes come from the original male inhabitants of the Cape. The history of oppression lives on in the genes.

The Cape Coloureds are also, in their diversity, a statement of the unifying power of sex. In the first days of apartheid, many South African families faced inspections in which they and their children were classified as White, Black or Coloured. The latter group posed a problem, for some children were light enough to pass for white, and were accepted as such, while others were dark and were classified as black. One official was sure that he 'could tell a Coloured from the way he spits' but even that was not always dependable. In the Senate's 1950 debate on the Population Registration Act (which set up formal racial classification) a Nationalist member was outraged by the admission of people of mixed ancestry into a superior category: '... we know that these people are Coloured, but because by repute and common consent they are white, we are going to make them White. By so doing we are going to allow Coloured blood into this race which we, some of us, wish to maintain so wonderfully pure.' In spite of his confidence, Afrikaners themselves have plenty of non-white blood, for a typical member of that group – the instigators of the apartheid regime – has about a twentieth of his or her DNA from African or Asian sources.

That episode was shameful, but has a lesson for biology. It shows how sex blurs history and brings families together as it mixes inherited information from separate lineages. The habit brought democracy to South African DNA, if not to its bearers. Over the centuries, the reproductive habits of Khoi-San, Europeans and Bantu spread genes through a whole community. The authorities – who, like the authors of Genesis, saw descent as a linear process – were often baffled by the appearance of what seemed to be a white child to coloured parents or vice-versa. Sex was to blame.

That pastime generates a web, rather than a tree or even a ladder, of kinship that, quite soon, draws the whole world into its embrace. It means that every individual pattern of descent can be traced not just from Adam or from Eve, but from untold millions of people.

The agenda of the ancestry companies is, as a result, more ambiguous than many of their clients realise. Any individual passes just half his or her DNA on to each child. After a dozen or so generations of sexual reproduction – well within the ambitions of some ancestor-hunters – any section of double helix hence has no more than a limited chance of a successful passage through the genetic labyrinth. As a result, large numbers of genuine antecedents will go unrecorded in a molecular pedigree and, perhaps even worse, some grand figure with a real historical tie to a person alive today may have, because of such dilution, almost no molecular link with him. Those who find a sense of nobility from their explorations of the past are often, in biological terms, deluded.

Sex leads ancestry hunters to deceive themselves in other ways. Often, they use a variant rare in one place but common in another as evidence of their own direct link with a distant (and distinguished) people. With the exception of Y and mitochondrial lineages (and they too can be ambiguous) that leads to self-delusion. The assumption is that of de Gobineau: that pure and homogeneous groups have mixed to make the modern world. Each supposed 'homeland' is identified by its having the highest incidence of a particular variant. Someone of that type from far away is then described as having Irish, or Berber, or Tibetan blood. Unfortunately, genetics does not work like that. My own blood group is B, which is rare in Britain with only one person in twenty bearing a copy. In Northern India as many as one in three is so blessed, but it would be foolish to say that I have 'Indian blood' (or even, as some might put it, 'the blood of the Mogul Emperors').

The fact that common variants tend to have rather independent geographical patterns also confuses the issue. As someone who has the 'positive' variant in the rhesus blood group system I am more like a typical Chinese (and hence perhaps descended from Confucius), all of whom are rhesus-positive, than I am to the people of Northern Spain who have plenty of rhesus-negative. Which bloodline do I belong to?

Biologists, unlike ancestor-hunters, are more concerned with statistical differences among the peoples of the globe than in the antecedents of one person or another. To them, the world pedigree is an almost impenetrable jungle in which each family tree shares both roots and branches with its neighbours. Ancestry is a forest not of pines but of mangroves.

Its tangled state is forced onto it by arithmetic. Everyone has two parents. Almost all of us have four grandparents, and many more can rejoice in eight great-grandparents or even sixteen great-great grandparents. In a perfectly sexual population the figure doubles each generation. That process cannot go on for long, because the numbers soon become enormous, the world runs out of ancestors and every pedigree is forced to connect to all the others.

Brothers and sisters share both parents and, as a result, half their heritage. Cousins share two of their four grandparents and have a one in eight degree of similarity, while second cousins have two of their eight great-grandparents in common, and so on. In an ideal sexual universe every sperm would always meet an egg with which it shared no ancestors at all, but the numbers game prevents that. The marriage of relatives – close or distant – hence makes any population less sexual than it would otherwise be.

In the West marriages among close kin are now rare, but in Victorian times were widespread. In Africa, the Middle East and parts of India, they still are. Partnerships between individuals closer than cousins are forbidden in the Koran, but South Indian Hindu and Sephardic Jewish communities welcome them and even allow the closer marriages of uncle and niece. Worldwide, a tenth of all nuptials are between partners closer than second cousins, and more than a billion live in countries where between a fifth and a half of all such ceremonies are between kin. In Pakistan two decades ago, two thirds of marriages were of this kind, and the process has gone on for so long that the average degree of relatedness of two random citizens is that of first cousins once removed (the kinship of children to their parents' first cousins).

Royalty takes such exclusivity to an extreme, for its very existence is defined by ancestry. Some among the nobility were convinced that they embodied the bloodline of a god.

They went to great lengths to ensure that the precious fluid was not polluted by plebeian corpuscles. DNA tests of mummies from the Egyptian Eighteenth Dynasty (who flourished from around 1400 BC) revealed the five-generation pedigree of Tutankhamun. He was the son of a liaison between brother and sister, and the DNA of two foetuses found in his tomb hints that he himself had sex with his half-sister.

Crowned heads are now less selective about who is invited into the marital bed. Even so, their families show how a few generations of close copulation can prune a pedigree. For a child of quite unrelated parents, six generations ago there would be sixty-four people on the tree. For the Bourbons, the rulers of Spain and Austria, the pool of acceptable mates was tiny, which meant that Alfonso XII of Spain, who reigned from 1874 to 1885, had because of repeated marriages to relatives not sixty-four but just six ancestors in that generation.

Victoria herself wed her cousin Albert. Nine of her children and twenty-six of her grandchildren married into noble families across the Continent, often in unions made to reduce the chances of conflict (that failed, for in 1914 Kaiser Wilhelm of Germany found himself at war with his cousin George V). The habit continues among her descendants. Queen Elizabeth and Prince Philip are, at the same time, second cousins once removed and third cousins by virtue of their joint descent from Victoria. In addition, Elizabeth is the second cousin of the King of Norway, and third to the monarchs of Denmark, Sweden, Spain, Belgium and Luxembourg. All of them descend from a Johan Willem Friso, Prince of Orange, who died in 1711. The marriage of Prince William – the future monarch – to a commoner, Kate Middleton, sparked off speculation about blue-collar blood in a blue-blooded line. There was certainly some, because the Duchess of Cambridge, as she became, is descended from miners and road-sweepers – but she also has ties with William Petty FitzMaurice, the 1st Marquess of Lansdowne, who served as prime minister in the 1780s. In addition, Kate and William are twelfth cousins once removed, with shared descent from Sir Thomas Leighton, an Elizabethan governor of Guernsey and Jersey.

When it comes to relatedness, royalty keeps the paperwork, but most of us do not bother. Even so, everyone bears an obvious clue of identity that allows them to work out the shape of their local family trees.

The number of surnames in relation to the number of people in a particular place shows how connected its lineages may be; a large population with few names suggests a closed community, a place with almost as many names as there are inhabitants indicates the opposite. A high frequency of marriage between two individuals with the same surname – Attenborough, perhaps – is a further hint that both partners descend from the same person in the recent, or not-so-recent, past. The technique works better for rare names, but the information is so easy to collect that even the Joneses tell part of the tale. Such figures show that the greatest numbers of links among pedigrees in Europe are in Paris and in Rome, while the isolated villages of Galicia, Murcia and elsewhere in Spain tend to keep sex within the community.

The double helix as a whole is itself no more than a vast and extended biochemical surname. On the large scale, the molecule is cut up and reordered each generation as sperm and egg are made, but over lengths of a few thousand, or even a few million, adjacent DNA letters that does not happen very often. As a result, the children of brother-sister incest tend to have long segments of DNA inherited in double and identical copy, one from each parent, for their version of the molecule has experienced just a single generation's worth of admixture. The children of cousins have, on average, shorter segments of paired material as the DNA from their common grandparents has gone through two rounds of sexual disruption. As the shared individuals on a pedigree recede further and further into the past sex continues to reshuffle the double helix so that fewer and fewer stretches are inherited in double copy.

That logic can be turned on its head. The extent to which each person's, or each population's, DNA is marked by long paired sections of identical sequence can be used to estimate how much marriage among relatives there has been and can provide a snapshot of how interconnected the local pedigrees must be.

The Adriatic islands of Brač, Hvar and Korčula are now tourist resorts but for many years their inhabitants were – like so many in the Balkans – wary of strangers. Church records show plenty of marriage within families over the centuries. Many locals still find their partners in their native villages, but others have become attached to husbands or wives from distant places. A scan along the DNA of those whose ancestors stayed at home shows that around a third of them bear many double copies of segments more than ten million letters long.

The children of their mainland relatives have fewer such sections, and of Croats in general (and of Europeans as a whole) fewer yet. A comparison of molecules with the islands' marriage registers shows a precise fit.

Britain has its own isolates; people separated by distance or by culture from their fellows. The Orkney Islanders have long seen themselves as distinct from their Scottish neighbours. Their surnames are different, with titles such Isbister, Kelday and Tulloch, rather than the 'Macs' common on the mainland. Their marriage records go back to the eighteenth century and reveal that few links were made between Orkney families and those of Scotland, ten miles across a stormy sea. Now the genes show that Orcadians have kept themselves separate for much longer. They have almost as many doubled-up and lengthy stretches of DNA as do the islanders of the remote Adriatic. A quarter of the natives have such paired sections more than ten million DNA bases long – a proportion fifty times greater than among Scots as a whole. Gales, cold and isolation have beaten their family trees out of shape.

The disasters of history have done the same across the globe. In Africa massive stretches of matched DNA are rare, as evidence that its pedigrees are highly intertwined. Mainland Europe has more evidence of recent shared descent. China has more again, as proof that two Chinese have a higher chance of shared descent from the same precursor who lived not long ago than do two Africans. Remote islands have even higher levels of common ancestry, perhaps because, as men and women moved across an empty world, numbers were so small that marriage to a relative was unavoidable.

Whatever the details, in the end, everyone on the planet – African, British, Chinese and more – belongs on the same pedigree. How long ago did their universal ancestors live? The calculations are full of guesswork but give an unexpected insight into the past.

For men and women considered separately, the task should be simple. To follow the Y back further and further into the past leads, as it must, to 'Adam', the great-grandfather of us all. The chromosome can identify the progenitor of a lineage (as in the Attenboroughs) or of a tribe, as in the descendants of the High Kings of Ireland. A whole continent may retain evidence of an ancient patriarch, for a certain version of the Y is carried by more than a hundred million men across Europe, perhaps through the efforts of an energetic farmer of long ago. Travel further back and we can, in principle, find the universal ancestor of every male alive today. Mitochondria can do just the same for Eve.

The existence of those two individuals is not in doubt, but their lives were not as simple as many imagine. Whenever and wherever Adam lived, he was surrounded by other men who had no idea that their lineages would disappear and he himself could have had no insight that his Y chromosome would, one day, be at the root of the world male tree. The same is true of his female equivalent.

The search for their homeland has gone on for centuries. The Good Book is precise in its description, for Eden is the source of four rivers, the Pison, 'which compasseth the whole land of Havilah', a place rich in gold; the Gihon, 'that compasseth the whole land of Ethiopia'; the Hiddekel (which may be the Tigris), 'that is it which goeth toward the east of Assyria'; and the Euphrates. Many attempts have been made to fit that to the real world. The Garden has been found in, *inter alia,* Iraq, Turkey, Egypt, Sweden, Sri Lanka, Mongolia, Florida, California, Missouri, and Ohio. Some enthusiasts place it at the North Pole, while General Gordon, of Khartoum fame, located it on the Seychelles island of Praslin, his evidence, in his biographer's words, being 'the remarkable similarity between the ripe fruit of the Coco de Mer, a gigantic palm tree, and Eve's *pudenda*'. In 1960, the editor of *The Flying Saucer Review* placed it on Mars, with the canals as the biblical rivers.

Its age is equally contentious. Archbishop Ussher used the lifespans of the patriarchs who descended from Adam to establish a date of 4004 BC. To work out the real birthdays of the primal couple is not quite as simple. It depends on guesses about mutation rate, about population size and about past patterns of movement. Even so, the genes, combined with well-dated fossils, have begun to tell the tale. A count of the inherited errors that have built up as mankind made his way across the world shows that the male tree reaches its root in central Africa around a hundred thousand years ago.

The birthday of Eve can be tracked down in the same way. She lived, those sums tell us, rather less than two hundred thousand years ago, well before Adam. The two can never have met, let alone have committed the first and perhaps least original of all sins: fruit, followed by sex.

The gap emerges from differences in the shape of the male and female family trees. The name Abraham means 'Father of a Multitude' and there was plenty of polygamy in his day. The Book of Exodus allows a husband to take several wives, as long as the first is properly cared for. David had at least five plus a retinue of concubines, while Solomon 'loved many strange women ... and he had seven hundred wives, princesses, and three hundred concubines'. His court was, no doubt, overrun by children.

For his sons to find their shared male ancestor they would have to go back no further than the noble ruler himself. There were, in contrast, hundreds of Solomonic mothers and to uncover their common female predecessor the daughters of his palace would have to delve much deeper into history.

If some males monopolise many females, other men must get less than their fair share. This means that the male population is, in effect, smaller than the female. That leads to a more rapid loss of Y chromosomes compared to that of mitochondria, and a shorter trip on the time machine to be introduced to Adam rather than to Eve. The present generation has, as a result, around 40 per cent more ancient great-grandmothers than it has great-grandfathers.

The Y and the mitochondrial lines are no more than twigs on the spreading giant of the world inheritance tree. We are linked to the past not just through those two exclusive pathways – fathers to grandfathers, and mothers to grandmothers – but through the billions of men and women who have slept together since history began. That habit unites the human race. A hunt for shared ancestry through male and female lineages considered together can zigzag through the generations, from son to mother to grandfather, or from daughter to father to grandmother. To do so unearths a common root far sooner than does the tedious plod to Adam or to Eve.

How far back must we go to find the most recent shared ancestor for – say – all Welsh people or all Japanese? And how much further is it to the last person from whom everyone alive today – Welsh, Japanese, Nigerian, or Papuan – can trace descent? The calculation demands even more guesses, some of them wild, about the size of populations, their rate of growth and the length of each generation than were needed for Adam and for Eve. The extent to which people tend to mate with relatives and their tendency to avoid other nationalities (as I discovered to my cost in the ten years I spent as a Welshman in Edinburgh) also raises barriers and pushes back the date. Most of all, nobody knows how much movement there was hundreds or thousands of years ago; and no more than a few people have to shift between continents to drag them into the same sexual net.

Speculative as they are, the results are a surprise. In a population of around a thousand people everyone is likely to share the same ancestor about ten generations – some three hundred years – ago. The figure goes up at a regular rate for larger groups, which means that almost all native Britons can trace descent from a single anonymous individual on these islands who lived in about the thirteenth century. On the global scale, universal common ancestry emerges no more than a hundred generations ago – well into the Old Testament era, perhaps around the destruction of the First Temple in about 600 BC, and long after Archbishop Ussher's estimate of the date when Adam met Eve.

Confucius himself lived at about the time of the destruction of that edifice. As he said, 'By nature, men are nearly alike: by practice, they get to be wide apart.' He was more right than he knew, for in political terms – by practice – people are even further apart than they were in his day, as millions face starvation while almost as many kill themselves with excess. Nature, in contrast, has drawn us closer together. Movement, migration, and the breakdown of social barriers have begun to unite the families of the world. The proportion of people who identify themselves as mixed race in Britain has almost doubled in the past decade and one household in eight contains members of different ethnic origins. For about half of the nation's children with an Afro-Caribbean parent the other parent is white, so that on these islands the pedigrees of two continents will soon merge. The process began long ago. Seven Yorkshiremen bear the surname Revis (after Rievaulx Abbey in the county). Each carries a Y chromosome that came from West Africa, perhaps in the eighteenth century. It has been joined by millions more. In Western populations as a whole, comparison of the incidence of long doubled-up sections of DNA in samples collected from around 1900 to the present day also suggests that the extent of close mating has much reduced over the past century.

Barriers to sexual relations among groups have not disappeared. In the United States black-white unions make up only one in sixty new marriages today, far fewer than in Britain – but even there the incidence has shot up from fewer than one in a thousand when Barack Obama's parents tied the knot some fifty years ago. Many others are based not on race, but on creed and clan. The genes prove that Indian castes, for example, have kept themselves separate for thousands of years, but even there the barriers have started to break down. As a result, one day, all families will become one. As they do, the time since the most recent common ancestor will draw ever nearer. Our global great-great-grandparents, the double helix tells us, are getting younger every year.

CHAPTER ONE

IN THE BEGINNING

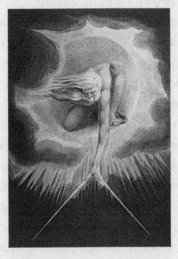

William Blake, *The Ancient of Days*

*In the beginning was the Word, and the Word was with
God, and the Word was God.*

GOSPEL ACCORDING TO ST JOHN 1:1

It began not with a word, but a bang. That statement is as
ambiguous as any in the Bible. The Psalms are confident
about what happened; the universe was the work of the Lord:
'Of old hast thou laid the foundation of the earth: and the
heavens are the work of thy hands.' That statement prompts

some obvious questions. When was 'of old'? What was there before the foundations were laid? And, most of all, what caused that sudden eruption into reality: did the event just happen, did it emerge from some mathematical improbability, or was it willed and if so by whom?

Such questions – of how time, the elements, life, and humankind find their origins – are at the roots of physics, of astronomy, of biology and, in a different sense, of belief itself. From the cosmos to the continents and from primeval slime to philosophy, everything evolves.

Science is an attempt to recover that process. The deeper it goes, the more indefinite its ideas tend to become but, for many of those who toil in its vineyards, obscurity increases the joy of the journey. As Sir Thomas Browne put it in *Urn Burial*, his 1658 speculation on human mortality: 'What Song the Syrens sang, or what name Achilles assumed when he hid himself among women, though puzzling Questions are not beyond all conjecture.' We may never hear the siren songs that coaxed the first stars, the first lives, or the first thoughts of the hereafter into existence but we can speculate about what they might have been and can now and again even come up with some evidence.

My own interests are in biology (and a small part of that subject, the genetics and evolution of snails – a topic that gets almost no further mention in these pages) but that discipline, like many others, rests on a foundation of chemistry and of physics. To put ourselves into true perspective most of this volume would have to be devoted to the Good Book's first three words: 'In the beginning ...' with the adventures of

Adam, Eve and their descendants reduced to a few lines in the final paragraph. Fortunately, I lack the knowledge (and the talent) needed to write even an abbreviated history of time. The Book of Genesis gets from the origin of the universe to that of *Homo sapiens* in fewer than seven hundred words. I cannot match its terseness, but this introductory chapter covers the same period at somewhat greater length. A small tale about a Big Bang and a rather larger one about the explosion of life is a preamble to the story of what makes us what we are. It reminds us that mankind lives in a minor solar system at the edge of a suburban galaxy, is in his physical frame scarcely distinguishable from the creatures that surround him, and – most of all – that he still understands rather little about his place in nature.

The universe itself, the product of that Bang, was once seen as proof that our home, and ourselves, were at the centre of everything. The Sun, Moon and stars were created to illuminate our ways: 'And let them be for lights in the firmament of the heaven to give light upon the earth: and it was so.' The cosmos was replete with theological lessons. A full moon looks flat, benign and almost supernatural. Medieval Christians saw it as a heavenly body, a jewel in the sky that in its divine exactness stood in stark contrast to the imperfect Earth, the realm of sin. In 1609 Galileo with his telescope put an end to that comforting thought, for he saw streaks of black thrown by mountain peaks in the lunar evening which showed that the Moon was a rugged world not much different from our own.

The truth about the origin of the Earth, and about when

and how 'the morning stars sang together, and all the sons of God shouted for joy' is more remarkable than the scribes imagined. The ancient paradox that the sky has stars rather than a universal blaze of light is proof that the cosmos expanded from a central point and left vast gaps between the shards of its first explosion. It must hence have a finite age. That claim is in Genesis, but in its modern guise was put forward by the Catholic priest and physicist Georges Lemaître. He called the birthplace of the universe the 'cosmic egg'. To ask what was there before that egg was laid is like asking what is north of the North Pole, for the question is based on a failure to understand the nature of time and space. In physics, as in philosophy, mere words can be deceptive.

At its inception the cosmos was dense, but is now dilute and as it expands becomes more so. When it set off on its journey is clear, for measurements of the relative positions of supernovae in relation to their age show that the famous Bang happened 13.77 billion years ago. The echoes of its explosion still sound through the universe. They warn of an uneasy past, a complex present and a gloomy future.

At the time of the Einsteinian revolution all radiation was thought to come from radium and its relatives from within the Earth itself. In 1910 the German physicist Theodor Wulf took an apparatus to measure electrical charge (that generated by ionising radiation included) to the top of the Eiffel Tower to compare levels three hundred metres up with those on the lawn below. They decreased, as expected, but less than predicted from measures made between two spots the same distance apart at ground level. He suggested that perhaps extra waves were streaming in from space. The idea was ridiculed but soon the Austrian Victor Hess ascended in a balloon to more than five thousand metres and found that, far from the radiation fading to zero as its source was choked off by the atmosphere, there was a three-fold increase in output, not much diminished when the Sun was blocked by the Moon in a total eclipse. He had discovered cosmic rays. That was the first hint that the universe was filled with energy, a relic of its violent conception.

At the very beginning there existed a quantum world of uncertainty ('a closed spherical space-time of zero radius') where the quantum effects of gravity dominated. Its enormous power resolved itself in a sudden expansion. We understand much of the history of the cosmos from around ten to the minus forty-three seconds, when elementary particles collided with each other to give matter and antimatter, to the present day. During the first few minutes after the formative moment, neutrons and protons combined to form hydrogen and helium, together with traces of deuterium and lithium, and within no more than a few million years the heavier elements emerged as those lighter particles fused together. What had been a fiery mass of infant chemistry began to congeal under the influence of gravity into the stars and galaxies that fly apart today. Now and again they implode into themselves or explode into the void.

What sparked off the Bang is a mystery. For believers, God did it; but to most scientists that statement is not an answer but an excuse. As those who study the skies struggle to fit mathematics to reality, some of their suggestions are almost beyond comprehension. Might there be an infinity of other universes out there, some of them precise copies of our own? Why does the rate of cosmic inflation seem to be on the increase, rather than slowing down under the influence of gravity? Where is the undiscovered nine-tenths of the universe, the 'dark matter' and 'dark energy', that might cause its expansion to accelerate or slow down? We do not know; and the explanations, when they emerge, will be at an intellectual pole that most of us will never reach.

Some ten billion years after the Bang the universe contained at least two hundred billion galaxies. Our own, the Milky Way, has about three hundred billion stars and perhaps twice that number of planets. Each solar system emerged when a cloud of molecules began to collapse in upon itself. Like a dancer who draws in her outstretched arms as she pirouettes it span faster and faster. Our own balletic debris condensed into the Sun – the 'greater light to rule the day' – and its eight planets, together with a confusion of rocky asteroids, distant lumps of ice, gas, and a cloud of dust. The Sun itself contains 99.9 per cent of the solar system's mass, most of it in the form of hydrogen, but is no more than a moderate object in the context of the universe as a whole. In an odd echo of the ancient notion of the Moon as a perfect heavenly crystal, the Sun is an almost precise sphere, and does not (as do many stars and the Earth itself) bulge at the equator because of the centrifugal force as it spins. Why, we do not know. The star is so dense that internal forces cause protons to join together through nuclear fusion, with a huge release of energy. Much of that is visible light, itself a player in the earliest pages of the biblical tale.

After its laconic account of the birth of the cosmos, Genesis moves on to an equally economical explanation of the shape of our own planet ('Let the waters under the heaven be gathered together unto one place, and let the dry land appear: and it was so'). That statement, too, has an echo in science.

The infant days of the Earth are known as the Hadean era. That geological hell is defined as the period before the first dated rock. It ended four billion years ago. Our planet was then an inferno. It rotated so fast that each day was no more than five or six hours long. The Moon ('the lesser light to rule the night') was formed when a body the size of today's Venus collided with another the size of Mars. A mass of vaporised minerals spilled out into space and the Moon condensed from this, perhaps as two separate objects which pursued separate careers until the one smacked into the other to make the present satellite, its far side quite different from its familiar face. The energy released melted the Earth's surface and sent clouds of dust into the atmosphere, which became fiercely hot. The Moon was far closer than today and friction caused by enormous tides generated even more heat, as did residual energy from the Earth itself as it cooled. In those torrid conditions gravity pulled the heaviest elements, iron most of all, towards the centre. There they still form a massive core.

A few reminders of that time remain. The Jack Hills in Western Australia are among the world's most ancient deposits. They contain layers of zircons, crystals of tough igneous rock, the oldest of which date from almost four and a half billion years ago. Their chemistry shows that they were made in the presence of liquid water as a hint that at least some of the ingredients needed for life were present long ago.

At about that time, the surface began to cool and solidify. Major traumas were still to come. Just under four billion years ago, a bombardment of objects hundreds of kilometres across hit our homeland. Their impact generated enough energy to boil the oceans. The atmosphere then contained carbon dioxide, nitrogen, water vapour and a little hydrogen. The Earth was – unlike some of its neighbours – massive enough to generate a gravitational field strong enough to keep hold of its gaseous cloak.

Our local star was then – as it still is – subject to internal upheavals that sent out great bursts of electrical energy. In 1859 (a year of solar storms as well as of *On the Origin of Species*) American telegraph operators were astonished to find that they could communicate across the continent with no need to plug in batteries. Energy from the Sun had excited thousands of miles of copper wire and generated enough electricity to do the job. Earlier tempests of this kind might have been powerful enough to strip Earth of most of its water and air (as, with the help of cosmic rays, they did on Mars) but its metallic core generated a magnetic field powerful enough to act as a shield.

Some of the 'waters under the heaven' emerged from local chemistry, but they were supplemented by a solid dose of the divine liquid from above the firmament. Many asteroids are covered with a thick layer of ice, and as they pounded the Earth they delivered billions of gallons of the precious fluid.

As the waters became divided from the land the first

continents appeared. Our planet differs from its neighbours in that – unlike those stolid celestial lumps – its skin is in constant upheaval. The Jack Hills zircons were ejected in pulses separated by fifty to a hundred million years, proof that in those primal days material was recycled from surface to crust and back again.

Even today the surface never stands still. Its movements are lubricated by water itself, for the landscapes of our arid neighbours the Moon, Mars and Venus are sterile and still. The ground upon which we stand is in constant motion as continental drift moves it on, breaks it up, and sucks it back into the depths.

In Australia I once saw a T-shirt that said 'Reunite Gondwanaland'; a demand that the world should rejoin the southern continent (or, perhaps, that Australia should join the world). Gondwana (as the place is more often called) was a five hundred million-year-old chunk of the Earth's crust that shattered to form parts of Australia, Africa, South America and India. It had an equivalent in the Northern Hemisphere called Laurasia, which became the raw material of much of modern North America, Europe and northern Asia.

Those two great landscapes were the latest in a long series of terrestrial reinventions. Australia itself has been chewed up, spat out, and glued together many times since the Jack Hills zircons came to the surface. Gondwana and Laurasia are relative newcomers. They split off a larger and older mass called Pangaea. Rodinia (after the Russian for motherland) was an even earlier player in the supercontinent stakes. It appeared as a great block around a billion years ago, and broke up some two hundred and fifty million years later. Much of it sank back underground to be melted and recycled, but evidence of its past lies in the roots of ancient mountain ranges scattered across the globe.

Two billion years before the Russian motherland came Ur, named after the ancient birthplace of Abraham. It may have been the first great land-mass of all, for the Jack Hills zircons suggest that in their day the crust was recycled far faster than at present and that, even earlier, when the rocks were still almost liquid, the surface, like porridge, bubbled and churned with no section permanent enough to be regarded as a continent. Ur itself broke apart and rejoined more than once before its demise and in most places has left almost no trace of its presence. Only in Australia are elements of the patriarch's homeland still on view. The modern continent has been assembled from several different landscapes, some ancient indeed. The sturdy natives of Adelaide or Perth have a history quite distinct from that of the effete inhabitants of New South Wales and Victoria. Their boots are planted on the remnants of Ur, while the sandals of Sydney skip over much more recent sediments.

Geological Arks, made of rock rather than the biblical gopher wood, have floated across the globe for untold millions of years. Earthquakes, volcanoes, tsunamis and the like are the bow-waves and wakes of continents as they travel across the globe and collide. The idea of continental drift explains the distribution of many creatures, from the flightless birds of Australia and South America that so puzzled Victorian biologists to the fossil beech forests now buried beneath Antarctic ice even as their brethren flourish in Africa. The geological lifeboats are still on the move. Gondwana itself will in time be reunited, for in two hundred and fifty million years another huge continent – Neopangaea, as it has been named – will emerge as nations skate across the map. Australia will clamber up the beaches of south-east Asia while the Atlantic and Mediterranean will be replaced by mountains. In years to come the waters will be parted from the land in a pattern quite unlike that of today.

Life itself was born in the ruins of such a slow geological car-crash. The uneasy movements of the newborn planet provided many of its ingredients for as the ground churned it dug up minerals from the depths. They were washed away by the rain to make a fecund chemical broth, the fare of the first organisms.

Astronomers understand much more about what took place just after the birth of the cosmos, when mathematics was transformed into physics and then into chemistry, than about why and how the Bang itself took place. The origin of living matter is much the same, for we know quite a lot about what happened soon after it emerged, but much less about how chemistry transmuted itself into biology. The search for an answer has sparked off dozens of ideas, some less improbable than others. What evidence exists comes from ancient geological deposits, from modern environments that might resemble those of long ago, and from experiments that try to mimic our homeland's first days. In addition, attempts to build simple organisms from scratch are under way. If they succeed life on Earth will be born again, which is of interest to biologists and might even attract the attention of a few theologians.

The biblical version has: 'And God said, Let the waters bring forth abundantly the moving creature that hath life, and fowl that may fly above the earth in the open firmament of heaven', which brings the birds in rather too early. In truth the vital spark appeared within a billion years after the origin of the Earth. As in Genesis, life burst into flame soon after the seas and the continents were formed. Every step, from simple chemistry to molecules that could copy themselves, from those to the closed structures called cells, and onwards to organisms and to species, marked an increase in complexity. Each leap forward was made from the ruins of the past. Every living frame resounds to the echoes of biology's own Big Bang.

In the seventeenth century, Sir Thomas Browne – of *Urn Burial* fame – mocked the popular notion that life was constantly generated from inanimate material (he was particularly unkind about the claim that mice emerged from mouldy hay). Even so, the idea persisted until Louis Pasteur did his famous experiment with sterile broth kept in a flask with an open arm bent in the form of an 'S'. No spores or eggs could fall in, and the liquid stayed pure and clean (proof of the failure of earlier attempts without the S-bend, which grew maggots as apparent evidence of the appearance of new forms). Pasteur saw the theological relevance of his result: 'What a victory for materialism if it could be affirmed that it rests on the established fact that matter organizes itself, takes on life itself; ... What good would be the idea of a Creator God?'

If life could not be made in a flask, where had it come from? There once reigned the notion of 'vital force', a descendant of the claim that, after Adam had been formed from the dust of the ground, the Lord 'breathed into his nostrils the breath of life'. It was a parallel to the 'ether' that astronomers then thought filled the universe. Matter seemed to exist in two forms; inorganic, such as gold or lead, which could be melted and would return to its original state when cooled, and organic substances such as chickens or chips which, once heated up, could never be restored. Their failure to get back to normal was, the experimenters assumed, because the material's *élan vital* had been driven off by heat. Organic materials could not be generated from inorganic, for the latter lacked the magic factor.

On that theory it should be impossible to synthesise the molecules of the body in the laboratory, but a German chemist soon did the job with urea, as proof that Leopold Bloom's tang of faintly scented urine as he ate a kidney was no more than chemistry. Soon, the idea that life needed an essence was abandoned.

We understand much less about the origin of biology than about that of physics and chemistry. We do not even know whether it was a unique event, or whether other forms may – like the multiple universes invoked by some physicists – have come and gone on Earth or elsewhere. Our immediate neighbours are not hospitable. This planet is the only member of the solar system to have liquid water and sufficient water vapour and carbon dioxide in the air to act as a greenhouse. The two gases, with the help of methane, raise global temperatures by fifteen degrees. Without them Earth would be a snowball. Most of the others fall at the first hurdle. Some have so dense an atmosphere that they are transformed into furnaces while others are so deficient that they freeze solid. Mars is an icy desert. As the Curiosity Rover showed in 2012 the place was long ago blessed with streams of water and may also once have had air but the planet suffers, in spite of its martial reputation, from diminutive dimensions and could not persuade its gaseous cloak to stay around. Perhaps there were Martians but they starved, froze, or gasped to death long ago.

One of the several moons on the outer edge of the solar system is almost half water, but only in the form of a glacier hundreds of kilometres thick. Others have the stuff in liquid form, but locked into the interior. Saturn's moon Enceladus has fountains that might generate the essential spirit but they show few signs of organic matter.

Living beings might also be present on some of the innumerable Earth-like objects in the universe as a whole. Sudden drops in the output of distant stars as their satellites pass in front of them have already led to the discovery of three thousand such planets, more than a thousand of which are about the same size as our own. A dozen can be defined, by their distance from their star, their size, and their possession of water, as potential homes for the vital spark. There must be far more. The Search for Extra-Terrestrial Intelligence programme is convinced that alien creatures have appeared somewhere and that they must of their nature evolve to get smarter. They might even send out signals to be picked up by their fellows. So far its radio telescopes have heard nothing more than noise.

Wherever it might begin, life needs raw materials. Where did they come from? In 1953 an American graduate student, Stanley Miller, appeared on the cover of *Time* magazine. He had made amino acids – the individual elements of proteins – by passing sparks through a mixture of gases held in a glass vial. His discovery was hailed as a new Creation. The conditions were rather like those rained upon Sodom and Gomorrah, with fire and brimstone (otherwise known as sulphur). Miller used methane, ammonia, hydrogen and steam, together with hydrogen sulphide and carbon dioxide. They might, he speculated, represent conditions close to those of the young Earth. Fifty years after his moment of fame, his samples were rediscovered. They were analysed with technology millions of times more sensitive than his own. The samples contained not just a few amino acids but twenty-three, in about the same proportions as are found in cells. Perhaps the young chemist had, after all, re-created the first moves towards the basic ingredients of existence.

Some of them may have been born far away. The Murchison meteorite is an object older than the Sun that crashed in Australia in 1969. Deep in its interior is a variety of compounds, picked up as it passed through the early solar system. Seventy are amino acids. They include all those made by Miller. There are also traces of the raw material of DNA itself. Some of the fuel of our being may hence have come, like the waters of the oceans, from a remote corner of the universe.

Space is like a cathedral; a large, cold, empty, overwhelming, and sometimes grimy place. It has plenty of carbon, the basic component of all biological molecules. High temperature and intense radiation persuade it to produce quite complex chemicals. Minute flakes of dust help for they act as centres of attraction for substances that otherwise might never meet as they float through immensity. In 2009 NASA's Stardust probe came back to Earth from its trip to distant comets. It had collected particles coated with the amino acid glycine, abundant in proteins. The universe has other unexpected talents. The Aquila galaxy, ten thousand light years from home, contains a cloud of ethyl alcohol big enough for a bottle of gin a day for every person on Earth for the next five thousand billion years. A successor to the Miller experiment hints at how that noble spirit may have helped us into existence. A blast of ultraviolet onto alcoholic stardust plus a dash of ammonia makes amino acids. If polarised UV, whose waves vibrate in a single plane, is used in the experiment, then the new molecules are twisted to one side – as are those in our own bodies.

Miller's results have become part of popular culture but science has moved on. His ideas were in the tradition of the gastronomic notion that existence evolved through a series of courses, from primeval soup to nuts. In fact soup, of whatever flavour, is dull and unadventurous stuff. The Miller model faces the Minestrone Dilemma; how was the leap made from a homogeneous ancient liquid to the lumps known as cells and organisms?

Wherever they came from, on the infant Earth the

ancestral molecules must have been rare indeed. Dissolved in a vast ocean, they would have almost no chance to meet a potential partner. To take the next step they needed support; a surface upon which to interact with their fellows. Our earliest predecessors may have been born on a bed of mineral particles to which molecules could bind. The Bible uses that idea as a metaphor: 'But now, O Lord, thou art our father; we are the clay, and thou our potter; and we all are the work of thy hand.' It was once thought that clay itself could act as the surface to which the first molecules could bind, but the chemistry of that does not seem to work.

The potter's wheel may have been made of another mineral. Serpentine is formed when rocks rich in iron and magnesium interact with water. The process generates hydrogen, methane and ammonia and generates a fine spongy mass, within the pores of which chemicals can – as they do on stardust – bind to a surface. That gives them a chance to combine with their neighbours.

Quite where the submarine reaction took place is not yet clear. Deep in the abyss, at the cracks that mark the birthplace of each continental plate, are tall chimneys that billow high-pressure hot water from the depths. They are the home of billions of bacteria and of other remarkable creatures. Their energy comes not from the Sun but from the remnants of the planet's fiery birth. The liquid that belches forth is laden with minerals and the rock it forms is porous and spongy. Poisonous gases such as carbon monoxide and hydrogen sulphide bubble from below and react with bound oxygen to make amino acids. Two hundred such places are known.

Their ancient equivalents have been hailed as a site for the first steps towards life, but the temperature is very high and the chemistry not much like that used to generate the relevant molecules in the laboratory.

A more promising candidate for the submarine Garden of Eden is known as a 'white smoker'. The name is rather misleading for such places do not draw vast quantities of energy from far below. The first was found a decade or so ago. The 'Lost City' stands near the summit of the Atlantis Massif, a submarine Alp off the coast of North Africa, not far from the boiling fountains of the Mid-Atlantic Ridge. It soars from the depths to reach seven hundred and fifty metres from the surface. The massif is built not of the black basalt that makes the floor of the great seas but stands on dense green serpentine that sprang from deep in the crust. The Lost City is a collection of monoliths that culminates in Poseidon, a tower sixty metres high. Hot water oozes from below, but at a temperature of just ninety degrees. Minerals leached out by this stream build the chimney walls. They are white because they contain compounds of calcium, barium and silicon. Methane and hydrogen leak from the surface. The rock is filled with fine pores that can bind simple chemicals and allow them to accumulate and to become interlinked. Many more such sites no doubt remain to be discovered. All the requirements needed to animate the inanimate are there. We may, perhaps, find our roots in Atlantis.

Whenever, wherever and however often biology's birthday was celebrated, just a single such event has left evidence today. All creatures that dwell on the Earth appear to descend from the same distant ancestor for the three-letter code of DNA is more or less universal. Like the cosmos, biology must have emerged suddenly, or not at all. The first few minutes, hours or days of a molecule able to take control of its own chemistry must have been precarious indeed. The forces behind its origin may, like those of the universe, have been quite different from those at work later. We may never know what they were. Biology is not like physics, for it destroys its own history. Natural selection – the machine that crafted it – eliminates those who came before. Life's ability to dig up its foundations probably erased the earliest forms at once and replaced them by more efficient creatures. If so, the primal moment has been lost for ever.

In 1943 the physicist Erwin Schrödinger gave a series of lectures in Dublin, later published as a book entitled *What is Life?* Every student of biology is posed that question at some time in his or her career, but their discussions tend to end in feeble generalities. I remember little about my own except that my fellow schoolboys decided that the talent needed food and sex (its products were also, as I recall, 'irritable'). Schrödinger had more insight. He saw that existence would need hereditary material and suggested that it might be a crystal; not a repeat of a simple unit as in diamond or salt but a structure able to modify itself as it was copied, to generate a variety of forms. His idea, James Watson and Francis Crick said later, directed them towards DNA.

The physicist also pointed at a more subtle demand, for order. All living things have an inside and an outside. They can pump stuff in and out. Within their walls they make their own environment, safe from the random chemical noise around them. Schrödinger's paradox is that biology is organised while everything else is not. His solution was to accept that every organism increases its level of internal orderliness at the expense of generating more chaos outside. The first step towards the crowded modern world may not have been molecules that could copy themselves, but membranes behind which they could find a safe place to carry out that ticklish task.

To keep itself in business the primal being needed to generate energy and to reproduce. Within its fortress, which came first, catalysts or copyists? Amino acids, and even simple proteins, are easy to make and can speed up reactions but they cannot generate new versions of themselves. The double helix, DNA, copes well with that challenge, but lacks the power to catalyse. Ribose nucleic acid, or RNA, does several jobs in the modern cell. It carries information from the DNA, plays a central part in the assembly of proteins, and regulates gene activity. It could be a compromise candidate as founder molecule for it can both replicate itself and act as a catalyst. Some of its constituents have been found in meteorites, and one of its ingredients is abundant in the interstellar gas.

Its units can also be persuaded to link into short chains if provided with a surface to which they can attach. In the laboratory RNA can be made to evolve at speed. If just the longest fragments in a flask are allowed to survive and are fed with the requisite raw material, molecules hundreds of units long appear in just a few hours. Short fragments of the stuff can even cooperate with each other to make further copies of themselves through a sort of positive feedback, in which several components work together much more efficiently than can one alone. Perhaps the first genes were not as selfish as they have been painted.

An artificial RNA with no more than five elements can promote a key step in the synthesis of proteins. Such a system may have fed upon itself to generate longer versions of both molecules. RNA is an imperfect scribe and the many errors made as it copies itself limit how long its molecules can become, which means that it now persists as the genetic material only in viruses. In time, DNA took over.

The double helix has an intimate relationship with the liquid that made biology possible. Long ago, Thales of Miletus, the first philosopher to try to explain the world without recourse to myth, came to much the same conclusion when he said that 'everything is water' and he was almost right. That banal fluid is remarkable stuff.

Almost all creatures, from bacteria to people, do best at around thirty-seven degrees Celsius. The average temperature of the Earth's surface is around fifteen degrees. Why do we like it so hot? Our preferred temperature is not far below the lethal level and most people find it hard to deal with a summer day not much warmer than their own bodies.

The hidden face of water is to blame. At body temperature, the molecule wraps itself around DNA to form a protective cage that keeps it in shape. Too cold and it cannot do the job, but just a few degrees warmer and the double helix shifts to a form that prefers to float free, as a single helix.

It then loses its liquid shield and collapses. DNA has a second precise relationship with the fluid that surrounds it. When unravelled into a single strand as it divides, it forms molecular bridges with water molecules. They guard the delicate structure from breakdown. Should one strand meet another the watery links are just weak enough to allow each helical molecule to disassociate from their embrace and to form a stable double structure. Water was the scaffold of life.

Schrödinger's tiny chemical universe lives behind a rampart, the cell membrane. Its constituents have the odd property that one end is at ease in water while the other is more at home in fat (soap is much the same). They hence make a double layer, with the fat-loving tails at the centre, and heads that protrude into the environment inside and outside. Such bipolar substances can be made in the laboratory in conditions rather like those of ancient times. The artificial sheets form globules which can grow and fuse with each other. They can even throw out strands which curl up to make a hollow sphere able to trap chemicals within.

A variety of pores, channels and sensory molecules are embedded within the membrane. They patrol the borders and pump charged particles in and out. Like a dam across a river, they hold in tiny units of positive charge until a reservoir large enough to drive a chemical reaction has been built up.

The jump from a closed sac filled with primitive replicators to a modern cell in all its complexity seems enormous, but has been bridged with the help of inherited differences in the ability to reproduce. Natural selection, as the process is called, makes, in its pedestrian and unplanned way, almost impossible things, from a primitive replicator to a human brain.

The powers of Darwin's architect may have been even greater in the days when an empty planet was ready to be exploited. A bacterium that weighs no more than a trillionth of a gram and able to copy itself within twenty minutes could make a ball of descendants the size of the Earth within two days. The fuse of life was lit in a dark ocean, and the ancient seas soon became busy and competitive places.

The chances that a relic of the earliest ancestors will ever be found are slim indeed. The chase for the first fossil can never end, but the latest candidates come from the Strelley Pool Formation in the Western Australian desert. Three and a half billion years ago the place was a warm and shallow sea. In a bed formed from a beach on one of the few islands that then dared to poke its head above the surface are preserved many cell-like structures, some arranged into chains and clusters. They may have been colonial bacteria.

Remnants of earlier organisms have not yet been found, but certain rocks have 'biosignatures' that hint of their presence. Greenland has areas of black shale – sedimentary rocks filled with carbon that may be of biological origin – almost four billion years old. They, like the stones of Strelley Pool, contain 'fool's gold', iron pyrites, a compound of iron and sulphur. This could be the chemical autograph of a creature from a time before the atmosphere had oxygen. It used sunlight and the oxygen bound in sulphate minerals to generate energy and made pyrite as a waste product. Such chemistry is quite sophisticated and must have had predecessors of its own. Free oxygen did not come until much later, when the primitive metabolic machine began to run in reverse and to use solar energy to generate organic molecules, with the vital gas as a by-product.

The ancient fossil record is so incomplete and its remnants have been so transformed by time that it contains almost no reliable clues about the first organisms. Another approach to Schrödinger's question is to attack the problem from the other end; to take a single-celled creature and cut out more and more until it can no longer function. To do so might hint at what the earliest forms would have needed.

Bacterial cells, primitive as we might regard them, are complex indeed. *Escherichia coli* – the standard experimental bacillus, present in vast numbers in our own bodies – has more than twenty-five million components, of almost two thousand distinct types (by way of comparison, a desktop computer has a thousand items packed into a container hundreds of millions of times bigger and, needless to say, cannot copy itself).

The smallest known genome belongs to a bacterium that lives inside the guts of cicadas. It manages with no more than two hundred genes. However, it has handed most of its functions over to its host and does not represent what a minimal creature might need in the outside world. Almost all other bacteria, from the ocean to the soil to hot springs, have more than fourteen hundred such elements and this may be a lower limit for most forms. The simplest independent organisms, with the smallest known cells, are called mycoplasmas. They have around eight hundred genes. A series of assaults makes it possible to destroy almost half of those and in the laboratory at least their bearers survive. Different bacteria seem to need different components, but the absolute core always includes those that store and copy information, together with others that build the cell membrane and yet more that pump energy, as a support of Schrödinger's ideas. The latest move is to try to build tiny genomes from scratch and to test how big they need to be to support growth and division when inserted into a cell whose own DNA has been removed.

The genetic material of the earliest cells, like that of today's bacteria, floated unprotected within the cell membrane and those creatures generated energy with their own, rather feeble, chemistry. That arrangement was not at all efficient. As a result, existence stagnated for a billion years. Modern bacteria are still pretty torpid for, in spite of all the damage they do and vital tasks they carry out, their cells remain rather uniform. Bacteria were the first conservatives.

To escape from that economic backwater needed a revolution. Evolution did not gain real impetus until the emergence of the eukaryotes, cells with nuclei, in which the genetic material is held within a membrane, protected from the hurly-burly of the energy machine. Most such cells are ten to a hundred thousand times bigger than a typical bacterium, and have a variety of internal structures absent from those simple organisms. Even the least elaborate are far more advanced than what had gone before.

The Entry of the Eukaryotes was the overture to the opera of advanced existence. It depended not on the slow trudge of natural selection, which does no more than tinker with the imperfect to make it somewhat less so, but on a single spectacular event; an agreement between two distinct kinds of ancient cell. One welcomed in the other, first perhaps as a meal, then as a servant, and later as a slave. The tree of life was reshaped by the promiscuous transfer of information between distant lineages. We are, to a large degree, what our earliest ancestors ate.

Bacteria are stuck in a rut because they cannot generate enough energy to impose a decent dose of order on their inner selves. One of their ancestors solved the problem by making a take-over bid for an independent power company. Its representative entered – willingly or otherwise – another cell. There, it multiplied and, in time, the host hijacked almost all its machinery leaving just that used in generating energy behind. Its descendants evolved into today's mitochondria, the place where much of the fuel of life is burned. Most of those structures' DNA (arranged, like that of bacteria, in a circle) has moved to the host's nucleus. Some eukaryotic cells contain as many as a hundred thousand mitochondria. Their efforts led to a massive leap in productivity and resulted in an explosion of innovation. In an independent but analogous arrangement another group of bacteria hijacked a second lineage that could soak up the Sun's rays and used it to make a chloroplast, the structure that turned the world green and gave it the oxygen that allowed the higher animals to evolve (Genesis has a different view for the Lord commanded 'Let the earth bring forth grass' even before he created the Sun).

The arrival of the nucleus marked the birth of the truly modern cell, with its instruction manual kept safely apart from the shop floor behind a rampart. It may have evolved as a defence against the wiles of the mitochondria, which bombarded their new home with parasitic DNA and with toxic by-products. The nucleus and the mitochondria each seem to have emerged just once. If those unique events were the door that led to advanced life, the chances of its existence elsewhere must be small indeed. SETI, the search for intelligent creatures out in space, may have its electronic ears cocked in vain.

The next great leap forward came around two billion years ago with the evolution of multicellular organisms. The process happened several times, for algae, plants, animals and fungi have each gone in for the habit of their own volition. All the partners must share the same genetic make-up to ensure that some do not pursue their own selfish interests at the expense of the others. Simple fusion of unrelated individuals would hence not work, but an experiment in which yeast cells in culture, each of which normally has a solitary existence but occasionally links up with another, were allowed to sink slowly to the bottom of a vessel and only those that got there first were allowed to reproduce found that clumps of cells emerged within just a hundred generations. Rather than individuals getting together to clamber onto a higher rung of the evolutionary ladder, perhaps cells failed to separate after cell division, just like most of those in our own bodies do today.

Whatever the details of each step, bacteria, plants and animals – like galaxies – flew apart as time went on. In physics, all tends to chaos. As biology tries to cope with that unfortunate truth it calls on natural selection to maintain its own order against each new challenge. Life is forced into expedient after expedient. Soon a vast array of forms emerges. Hundreds of millions – perhaps billions – of species have existed on Earth. Almost all were in the end defeated by the laws of physics and of chemistry and left no descendants.

DNA is a biological telescope that can be used to peer into the past. The niceties need not detain us long, for – as in the expansion of the universe after its first few seconds – the appearance of new species does little more than provide a huge series of examples of the same thing. This abbreviated account of existence, like that in Genesis, hence deals just with its first stages and with the origin of what the scribes saw as its ultimate artefact, ourselves.

To the cold eye of the rationalist, the diversity of modern plants and animals is no more than a rickety, unplanned and in truth rather dull edifice that rests on ancient foundations. Evolution is a series of successful mistakes – and a far greater number of failures – in the endless battle against the world outside. Its tactics, once its machinery lumbers into action, seem in retrospect almost inevitable.

Inevitable or not, that process connects each reader of this book with the first cells, and with every other creature today. The double helix rebuilds the global family tree. Humans find their own place among the mammals (a group united by hair and milk). Their molecular pedigree links mammalian evolution to the break-up of Pangaea, the ancient world continent. Elephants, hyraxes, sea cows and golden moles, different as they now appear, trace a common root in Africa. Another group of mammals, the anteaters, tree sloths and armadillos, find their origin in what is now South America. The great northern continent of Laurasia was the birthplace of bats, whales, dogs, cats, hedgehogs and more, all of which were born on that now shattered landscape. Parts of Laurasia broke off to form islands which later emerged as North America, and as Europe, Siberia and parts of China. Rodents, hares and rabbits, trees shrews and – in the end – the primates, the group to which apes, monkeys, lemurs and humans belong, were all born on the island of Eurasia.

DNA places the primates next to the colugos, squirrel-sized animals of south-east Asian forests notable for flaps of skin between their front and hind limbs that allow them to glide from tree to tree. They come out at night to feed on leaves and fruit, and spend their days suspended beneath branches, where they cower in fear of eagles.

We have three hundred or so more immediate relatives among the primates as a whole. The group emerged about ninety million years ago. The hominids – humans, chimps, gorillas, orang-utans and gibbons – set off with an identity of their own some twenty million years before the present. The line that led to *Homo sapiens* diverged from that of chimps around eight million years ago or perhaps even longer, while gorillas took their separate way a million years before that.

Chimp, gorilla and human DNA has been read from end to end. Chimps and humans diverge in sequence by around 4 per cent. About two thirds of the difference comes from variation in the numbers of copies of various repeated gene sequences, with the rest due to shifts in individual letters of the code. There are also differences in the activities of certain genes, and in how the whole system is wired together. It has become fashionable to emphasise the similarities between ourselves and our closest relative but – given that a shift in just a single letter of the code can cause severe disease and that we diverge from chimps by tens of millions of such changes – the disparities are quite substantial.

Chimpanzees and humans are distinct in many ways. Our females, unlike theirs, have a hymen but our males lack a penis bone. Human kidneys retain less salt than those of chimps, but men and women can weep, float, dive and catch fish, all talents that our closest relative lacks. Most of us are right-handed, but about half of all chimps favour the left. We have chins and eyebrows and show the whites of our eyes while they do not. Chimps, on the other hand, are blessed with body hair and with jaw muscles much more powerful than our own. In addition, our infants are helpless to an advanced age and – unlike any other animal – we are acutely aware of the past and of the future (our own demise included) and are blessed with art, music, belief and above all speech.

We can infer some of our history from the remnants of the past. In the last half-century anthropologists have entered a valley of bones; not quite as populated as that of Ezekiel, but far better than those known even a few decades ago. In Darwin's day almost no human fossils had been recognised and even half a century later there were scarcely enough to cover a large laboratory table. Now, hundreds of relics of our ancestors and relatives have been found. The story of our bodies – and our minds – has begun to emerge from the rocks.

The remains tell a somewhat tangled tale and we have

almost no evidence that any of the fossils named as members of the human lineage have a direct link with anyone alive today. Some of the species may have been separate branches on a complicated pedigree. They shared ancestors, but not descendants, with ourselves. Whatever the details, just a single lineage has survived; the one to which we all belong.

For much of the past, in contrast, we shared the planet with creatures much closer to ourselves than are chimpanzees. Neanderthals, those supposedly brutish beasts, split from our own line around six hundred thousand years ago and occupied much of western Europe, although at times they travelled as far east as Siberia and as far south as the Middle East. They emerged in Europe and in western Asia from an ancestor who had escaped long before from Africa. With a stocky body and elongated face Neanderthal was a cold-adapted creature. The creature's entire genome has been read off from fossil bones. It diverges from that of modern humans by less than one part in a hundred. By forty thousand years ago, the Neanderthals were in retreat. Soon, no more than remnants were left, in Iberia and in the Balkans. Fossil sites and stone tools suggest that our own ancestors simply pushed them out.

Another group, the Denisovans, is known only from a forty thousand-year-old finger, toe and tooth in a cave in Siberia, a region in which both Neanderthals and our own ancestors lived. Fossil DNA shows that the Denisovans, too, were distinct. They were close kin to Neanderthals but their ancestors left Africa eight hundred thousand years before ours. So similar are their genomes and those of the Neanderthals to our own that a few of their genes might live on in peoples outside Africa, possibly through hybridisation with our direct ancestors in the distant past. The tiny and enigmatic creatures found on the island of Flores that lasted from about the time of the escape from Africa to no more than seventeen thousand years ago may be yet another branch of the recent human lineage.

All those creatures (except the last) were notable for the contents of their skulls. Weight for weight, the human brain is five times bigger than that of the chimpanzee and the speculative bits are larger yet. How did that happen?

When it comes to the origin of thought, grand theories rest on few facts. No doubt a whole complex of changes led to our giant dose of grey matter. One involved a real shift away from the other primates; the ability to stay upright for long periods. To do so offered a new way of life and demanded a leap in mental ability.

I cannot ride a bicycle because I spent the time when I should have been learning how in the local library. I envy those who can, but am more impressed by my simple capacity to stand on two legs. That is what really makes us human, for no other mammal can do the job for more than a few minutes. Even those with a youth as joyless as my own manage it with no trouble at all.

It was once assumed that the intellect led from the front; that an already clever primate stood up when apes had become, in some senses, men. However, when the 3.2-million-year-old *Australopithecus*, Lucy, was found, her remains, which included a skull, a pelvis, a spine and a leg, showed that she too could attain the vertical even though she had a small brain. Fossil footprints at Laetoli hint that the ability to stand had emerged even a million years earlier in a primate with a brain no larger than a chimpanzee. Living an upright life did not, it appeared, demand much of an intellectual effort. Walking seemed to have emerged before thinking.

The latest discoveries hint that it was harder to stand up than once was thought. A new *Australopithecus* fossil shows what its shoulder looked like. The shoulder-blades of apes have an upward-facing socket into which the arm-bones fit and the blade itself is at an oblique angle to the body, allowing the arms to rotate with ease, which is a help when swinging from branch to branch. Children, in contrast, have a downward-facing socket that points out to the side as they grow. That restricts the extent to which we can move our arms and keeps us firmly out of the trees.

Lucy and her fellows had chimp-like shoulders. The creatures must have spent plenty of time up in the branches where, perhaps, the mental challenges were less severe than on the ground. The first shoulder that looks human is from the Turkana Boy, who walked the Earth no more than 1.8 million years ago. Big brains did not come in feet first, but had to wait until the arms could also play a part.

An upright way of life calls for shifts in both body and mind. The foot, with its twenty or so bones, had to rearrange itself, while the pelvis, the spine, the arms, the chest and the neck, with the skull perched high and the eyes set to scan the horizon in the search for food, enemies, or mates were also much modified. Those who stood up paid a price. The chimp spine is stiff compared to ours and the animal cannot bend backwards, but the human backbone suffers more wear and tear, and dooms its elderly owners to arthritis together with slipped discs. Chimps have flat feet rather than our own elegant arch, but we pay for the latter in the currency of bunions and sprained ankles. Our knees and hips bear the whole load of an upright body and have grown larger to cope. Natural selection expanded the spongy inner bone and kept the thin, hard outer layer – which means weaker bones and real problems if we fall over in old age.

Given such difficulties, why the move from four limbs to two? There, guesswork reigns. Perhaps the rains failed and forest was replaced by open countryside, so that it became impossible to clamber from tree to tree. Men and women use no more than a quarter as much energy when on the ground as do chimpanzees. They can as a result travel much further and faster. All this allowed them to range further in search of food and to carry their finds back to a place where it could be shared.

A pelvis specialised for upright travel has a rather narrow birth canal so that babies must be born earlier in development than had previously been the case. That demands more interaction between mother and child. As the infants became less able to grasp fur with feet as well as hands, their mothers had to hold them tighter than in the days of the tree-dwellers. Perhaps women became less independent as a result.

As so often when it comes to speculation about the past, sex rears its hypothetical head. An upright stance might improve a man's ability to impress local females (and to depress nearby males) both directly and with an open view of the size of his penis. Punches hit harder when thrown from an upright position rather than from a crouch and this too may have caused angry or jealous males to reach for the sky. The same strategy might have frightened off rivals, or lions; and – unlike chimpanzees – we can clench our fists for a harder blow.

A different set of ideas has to do with the need to cope with the layer of super-heated air near the ground on a hot African day. The escape was to rear upwards, to breathe in a cooler breeze and to cut down the area exposed to the direct rays of the sun. Our naked and sweaty bodies may have evolved for the same reason. No chimpanzee could win a marathon for we have an unmatched ability to get rid of heat when running – which was hard luck for prey that could be chased for hours across the savannah until they collapsed.

Long legs may have led to a bigger brain in another way, for prolonged hard exercise – that uniquely human talent – causes muscles and nerves to release a growth factor into the blood which increases the number of connections between brain cells and which, when in reduced supply, is associated with Alzheimer's disease. A larger and better connected dose of grey matter provided a new inquisitiveness about the world. In time, that new-found gift extended to interest about the past and concern for the future.

The seeds of curiosity may hence have been sown long before our own species emerged. They came to fruition first in spirituality, and then in science. Both, in their own way, try to untangle the past and, with rather less success, speculate about what is to come.

The Bible's last book, Revelation, was written on the Greek island of Patmos, where John received a series of visions (Thomas Jefferson referred to its verses as 'the ravings of a maniac'). It has beasts that rise from the sea and ascend from the bottomless pit, four horsemen, a seal branded on the forehead of those who are to be saved, and more. A star called Wormwood falls from heaven and kills a multitude. In a series of bizarre images, the Book uses mathematics to reveal the plans of God, with repeated mentions of the sacred numbers three, seven and twelve.

Physics, with its own insight into how the cosmos began, can itself make statements about the end-times, albeit with rather less style. Like all stars, the Sun has its death foretold. Each year it grows brighter but it will, in the fullness of time, lose its power, swell, and engulf the Earth. Long before that, all life will come to an end as the source of the energy that allows it to keep physics and chemistry at bay disappears. In time, the whole universe will run down. Its energy gradients will even themselves out and an era of random noise will begin as gravity disappears. Whether the new cosmos will burn or freeze is still a matter of dispute among astronomers but St Peter was confident in his predictions: 'Looking for and hasting unto the coming of the day of God, wherein the heavens being on fire shall be dissolved, and the elements shall melt with fervent heat.' Whatever the climate, around ten followed by a hundred zeros years from now all will be darkness; and time, will, in effect, be at an end.

Bleak as that prospect might be, a few theoreticians suggest instead that time and fate are cyclical and that the universe will start again from scratch. The 'Big Crunch' theory claims that expansion will, in the distant future, go into reverse and the cosmos will collapse into a tiny point, only to expand anew, perhaps to repeat the process for ever. Man, or something like him, may then be born again in a new and perhaps less imperfect form. Should that happen, St John the Divine will have reason to be pleased.

will start again from scratch. The Big Crunch theory claims that expansion will, in the distant future, go into reverse and the cosmos will collapse into a tiny point, only to expand anew, perhaps to repeat the process for ever. Man, or something like him, may then be born again in a new and perhaps less imperfect form. Should that happen, St John the Divine will have reason to be pleased.

CHAPTER TWO

THE PLAYING FIELDS OF EDEN

William Blake, *Good and Evil Angels Struggling for Possession of a Child*

The race is not to the swift, nor the battle to the strong.

ECCLESIASTES 9:11

Type the phrase 'scientists find the gene for' into Google and thirty thousand results appear. Many are repeats and some relate to animals, but most refer to ourselves, which is impressive given that we have no more than twenty-three thousand genes (as traditionally defined) altogether. The hits include sections of DNA that cause illnesses such as cystic fibrosis or muscular dystrophy, but also others said to lie behind insomnia, premature ejaculation, musicality and marital failure. Within the mix lurk claims of biological controls alleged to act on a vast diversity of attributes, from happiness to homosexuality and from compassion to crime.

More and more, biologists claim to be able to read the book of life, the fate inscribed within the double helix. To do so might, some say, distinguish those destined for physical or social salvation from those condemned to the opposite. The human genome – which took fifteen years to decipher for the first time – can now be sequenced in a few hours, for a few hundred pounds. Dozens of genetic diseases can be detected in a foetus or an infant and the information used either in treatment or to terminate a pregnancy. Some find that alarming, but will there soon also be probes to identify the athletic, the clever or the musical? Perhaps more important, what about the prisons and the workhouses? Some people may plunge into poverty or wrongdoing through their own innate weaknesses, struggle against them as they might. How will society cope? Genetics has been hailed as a salvation for the afflicted, but it may instead end up as a molecular Inquisition that picks out those programmed to fail.

As Ecclesiastes points out, the fastest are not always the first across the finishing line, nor do the beefiest invariably triumph in the ring. An inborn talent, however potent, is open to the circumstances in which it finds itself. Genes alone do not predict outcomes: the biblical verse quoted at the head of this chapter continues as '... neither yet bread to the wise, nor yet riches to men of understanding, nor yet favour to men of skill; but time and chance happeneth to them all.' Who triumphs and who fails turns on circumstances – on time and chance – as much as on biology. Nature and nurture work together to determine who wins or loses in the race through life.

Theologians have argued for years about such issues. The notion of inborn moral frailty and what to do about it pervades Christian teaching, from the grim certainties of damnation held by Calvin and his Reformation fellows to the weaker and more forgiving pieties of today. Since Adam and Eve accepted the promised chance of knowledge of good and evil, their descendants have been forced to pay the price in the form of original sin. To biblical literalists, on Judgement Day will come a final examination in which everyone's strengths and imperfections will be revealed and their destination, be it Heaven or Hell, determined by the Lord himself. Only the free acceptance of the Gospel message offers hope. The quandary is that of the tension between nature and nurture; between native weakness and the free decision to opt for salvation.

The notion of destiny pervades theology and biology alike. Is humankind's fate inborn, or is it formed by the circumstances in which we live? How will men and women, flawed as they might be, cope with their own weaknesses? And what if their prospects can be revealed at, or even before, birth?

Such questions are behind much of the public's concern about the new powers of genetics: are people born with a fate coded into the double helix and, if they are, how should society cope with it? Should those found to be imperfect be forgiven for their failings, or should they be punished? And what of the naturally talented – should they be left to their own devices, be given special treatment, or penalised to give others a chance?

From sports field to schoolroom and from pulpit to prison the experts do not agree. In truth, much of the debate lacks meaning, for nature and nurture are so intertwined that they usually cannot be separated. That fact is forgotten by most of the general public, by many theologians, and sometimes by biologists themselves.

Sport – Ecclesiastes' metaphor for the race through life – is a useful microcosm of the wider dispute about the role of inborn talent. It shows how the response to the new genetics by educationalists, doctors, lawyers, bookmakers or, for that matter, theologians is confused and inconsistent. It also demonstrates, better than almost anything else, how hard it is to separate inheritance from experience, how differences that are unimportant in some circumstances are crucial in others and how the role of nature depends absolutely on what nurture has been encountered.

The power – and the problem – of prior knowledge in sport become very clear when making bets on the result. A punter at the Derby needs to know a lot about form: about the past successes and failures of the day's runners together with as much as he can discover about their family history and how well they have been trained. Bookmakers use the same evidence to set their odds. They can for most of the time guarantee themselves a profit, for they are almost always better informed than are their customers. The transaction between the two parties is a battle of wits, confused by the random noise that emerges from the condition of the track and from the mental and physical state of horse, jockey, turf accountant and gambler on the day.

The authorities do not hesitate to use information on individual talent to level the playing field, for in half the meetings held in Britain a likely winner – a steed with a good record – is obliged to pay a penalty to slow it down. Without such a handicap, the Sport of Kings would become dull and in time the gaming industry would collapse, for the same nags would almost always be first past the post. To sharpen the odds further (and, as an incidental, to enlarge their market) those in charge also ensure that within any event horses are matched against each other in terms of quality – plodder versus plodder, or champion against champion – for that, too, increases the numbers of contests with a less than certain outcome.

The handicapper's job is not simple. A dozen officials go through the data and publish a list each week, based on each animal's record, its age, its sex, the interval since its most recent race, the month in which the event is held, and more. All this information is weighed up and the decision made as to what level of excellence the horse has reached. The figures are translated into 'imposts', strips of lead placed into the saddles to give the less able a chance against their more talented rivals. A similar procedure (without the lead strips) is also applied to activities as different as polo, basketball, golf, croquet, boxing and even chess, with the most accomplished faced with penalties to give the less well-endowed some hope. To sort horses, or football players, into divisions of different quality generates huge sums as investors buy the best in their attempts to ensure that their team, or stud, stays in a high league.

In other sports such naked attempts at bias are unknown. Every contestant competes as an equal and, depending on the nature of the contest, the fastest, the strongest, the most brutal or the most graceful takes the crown. If he or she has won before, nobody cares. It would be bizarre to force a champion sprinter such as Usain Bolt to carry a crate of beer in the two hundred metres just to give somebody else a chance. His success depends on his innate abilities together with a powerful sense of dedication. The Jamaican's record is so spectacular that those who bet on such events would receive poor odds in any contest in which he takes part.

Sometimes, the clues about likely winners and losers come from biology. Tall basketball players are at a premium as they reach for the hoop, but short gymnasts do better for they have a low centre of gravity while sumo wrestlers are marked by the survival of the fattest. I have hated team sports since my schooldays and made that clear whenever I was forced to participate. In sardonic mood, the games teacher once obliged me to put the shot in a competition. I came last, which, given my skinny frame and modest stature, was predictable but less than fair. My innate flaws led me to fail, and even had I been desperate to win my hopes would have been doomed by the implacable truths of inheritance.

When it comes to the propulsion of heavy steel balls many are called but few are chosen. The same is true for music, for art, for higher mathematics and perhaps even for entry into prison (or, for that matter, heaven). For such endeavours each of us is gifted or otherwise in our own way. Many people are convinced that the real experts have been born with a God-given ability. Others disagree. Daniel Barenboim, whose musical genius was obvious when he was five, was once asked whether he had ever come across any child prodigies whose brilliance matched his own. He replied: 'No, but I have met plenty of their parents.'

Nobody denies that Barenboim's virtuosity comes in part from years of practice but in spite of his cynical remark inheritance may indeed play a part in his abilities and in those of many others.

Genetics promises to give us the tools to study its role, not just in sport or musicality but in more subtle aspects of body and mind. It may make it easier to discover potential champions – the Barenboims of the future – at an early age. That may be gratifying for their parents, but might it not remove some of the uncertainty that is the spice of life? And what about the losers; not just those who come last on the sports field or in the piano competition but in the wider sense? Each of us is flawed or talented in one way or another, and if technology can predict every infant's chances of success or failure society may find it hard to cope. Should we penalise those born with a propensity for crime, or forgive them? Should more money be spent on the education of clever children, or would the cash be better spent on the dim, where it might do more good? Who do we handicap, and in whom should we invest?

Christianity has long been riven by debate about that issue. For some, the best hope of salvation is to pursue a righteous life. The fourth-century theologian Pelagius felt that the mark of sin had been wiped clean for all mankind by the death of the Saviour and that man could hence choose good or evil by his own volition. He had free will, and should use it for worthy ends. 'Obedience,' he said, 'results from a decision of the mind, not the substance of the body.' To live well was a certificate of readiness for the world to come.

Others rejected such an easy escape. They believed in 'total depravity'; the idea that Nature is all, and that individual acts, virtuous or otherwise, will not change it. That notion was embodied into the stark doctrines of the Reformation. Calvin and his followers based their logic on Paul's letter to the Ephesians: 'For by grace are ye saved through faith; and that not of yourselves: it is the gift of God: Not of works.' The winner of the divine Derby, the reformers insisted, was decided long before the race begins. Few would find themselves among the elect at the apocalyptic finishing post, for the Lord had defined large numbers of his subjects as 'filled with all unrighteousness, fornication, wickedness, covetousness, maliciousness; full of envy, murder, debate, deceit, malignity; whisperers, backbiters, haters of God, despiteful, proud, boasters, inventors of evil things, disobedient to parents, without understanding, covenant-breakers, without natural affection, implacable, unmerciful' (which covers most of us). No attempt to overcome those inborn flaws could ever succeed.

That stringent view was welcomed north of the Tweed. The Westminster Confession of Faith was drawn up during the English Civil War as a compromise between the Church of England and the austere canon of the Scots Presbyterians: 'By the decree of God, for the manifestation of his glory, some men and angels are predestinated unto everlasting life, and others foreordained to everlasting death.' Man is what God made him, and any attempt to avoid that truth is futile.

From birth onwards, the Lord knows who will, and who will not, win the race for salvation. Our own actions are beside the point. Some took the notion of predestination to the extreme, and in parts of Germany took up orgies and polygamy as a statement of their belief that earthly acts had no influence on the chances of salvation, which were set long before birth. The same philosophy (with fewer orgies) was adopted by the Puritans of Massachusetts.

The Roman Church looked askance at both Pelagius and predestination, for each sapped at its own powers. Man may have been poisoned by the toxic fruit of the tree of knowledge but the Church had the antidote. To obtain it, St Augustine and others insisted that Christians must recognise Rome's power to help absolve their sins (and if they refused, benignant asperity – the stake – was always available).

In the biological context, for some (and perhaps most) people Pelagius is right and a healthy and contented life will allow them to escape any inborn frailties. Others need the attentions of Augustine; they have innate flaws, but with the help of those competent to treat them, may be saved. A few, though, are unlucky indeed for their birthright is set in Calvinistic stone and their damnation, in terms of health or otherwise, is assured, much as they might try to avoid it.

The biology of fate is much discussed, particularly by non-biologists. In truth, the subject is far less developed than many people imagine and has become more (rather than less) difficult to interpret as technology progresses. The happy days of simple Mendelism – of common genes for common diseases, or even for common social problems – are long gone.

From its earliest days genetics has, like the Church, prom-ised rather more than it can deliver. Both enterprises face the lure of the obvious; the perception that processes as familiar as inheritance or destiny must be simple. They are not. Genes are not, as once assumed, simple beads on a string, but instead are broken up, subdivided, and separated by vast quantities of what was until not long ago described as junk but is now known to be the site of an intricate set of control mechanisms. Humans have, for reasons obscure, fewer sections that code for proteins than do tomatoes. When it comes to inherited dis-ease, similar symptoms may be caused by damage to different parts of the DNA, while the same error can give rise to what seem to be quite unrelated illnesses. The influence of the double helix on man's destiny is far more ambiguous than it once seemed reasonable to assume.

The hunt for the variants that might lie behind human diversity in height or happiness has run into another unex-pected problem; what might be called, after T. S. Eliot's feline master-criminal, Macavity genes. Again and again, the evi-dence that a particular characteristic runs in families is persuasive, but the agents responsible cannot be found. The hidden paw strikes again and again but more often than not, as scientists comb the double helix for clues, Macavity's not there. The work of the biological detective has turned out to be far harder than expected.

The height of children is so similar to that of their parents that large parts of the variation found in all populations must be associated with genetic factors. Stature is easy to measure, does not change much with age after adolescence, and is not laden with social or political import. Certain mutations have a dramatic effect on their bearers' inches, as the Irish Giant shows. Height seems to be an ideal system in which to find genes.

It has not been so simple. The molecular tape measure has been used on tens of thousands of people. In Europeans, two hundred different DNA variants are associated with differences in stature – but together they account for less than one part in ten of the diversity needed to explain the similarities of successive generations. A few are within, or near, genes that when they go wrong cause growth disorders, but most have no known function. Africans share some of the variants that predispose Europeans to being tall or short, but also have plenty of their own. Height is a gamble with huge numbers of cards, whose identity shifts from one lineage to the next. Each of us reaches our ordained dimensions in our own way and the notion that just a few genes determine what size we will reach is wrong.

Feet and inches are not of much practical importance (even if tall people tend to be more intelligent, richer and – some say – more attractive than those closer to average). Other attributes are. Cancer, heart disease and stroke are responsible for almost half of all deaths in the developed world. They run to some degree in families and several genes that place a very few people at high risk are known. Once again, though, most of the individual variants are rare. To complicate matters further, some people who bear variants known to predispose to illness remain perfectly healthy and some who are spared from such things develop the disease.

As technology advances, the news has got worse. The protein-coding sections of the DNA of many thousands of people with heart disease, high blood pressure, or diabetes, has now been scanned, but few links of any gene with such medical problems have been found. They may exist, but pessimists (or realists) now suggest that it may be necessary to survey hundreds of thousands of people to find them.

For a few simple diseases the ability to diagnose a foetus carrying a dangerous variant has been important for it has led to pregnancy termination and to a rapid drop in incidence. In Cyprus, thirty years ago, about one baby in a hundred and fifty was born with a crippling blood disease, at enormous social and financial cost to the patients, to their families, and to the state. Then, a compulsory screening programme for carriers was introduced and the illness has almost disappeared.

Nobody can deny the importance of such advances, but they are exceptions. The notion that every infant will be given a genetic passport at birth that lays out their medical (let alone their social) prospects until their dying day is seen by some as seductive, by others as horrific, but for most of us is irrelevant. A survey of tens of thousands of identical twins explored the extent to which they share conditions such as cancer, heart disease, diabetes, multiple sclerosis and several other conditions. For more than half the pairs there was sufficient difference between each twin to suggest that even a complete DNA sequence would have almost no predictive value. A few cases of thyroid disease, childhood diabetes, and Alzheimer's might be picked up before symptoms manifest themselves but the widespread claim that the new genetics will revolutionise diagnosis, let alone treatment, is premature. The hubris which accompanied the Human Genome Project has stalled.

Even when the science moves on again, as no doubt it will, genetics will always face a central issue: that whatever the role of inherited variants in the control of height or health, of sporting or intellectual talent, of worldly success or of abject failure, nature and nurture always work together. What genes can achieve depends on the environment in which they find themselves.

Across the world average height has much increased over the past few decades, not because of any changes in DNA but because of improvements in well-being. Young South Koreans are four inches taller than their kinfolk in the north in spite of the fact that they share almost the same genes. North Koreans are short because many have faced near-starvation in childhood. In much the same way, today's toxic Western diet of fat and sugar has doomed many Britons who would once have stayed healthy to heart attacks because their biology cannot cope. From athletics to obesity and from genius to crime, the more we learn about genes the more important the environment appears to be. Quite often, Nature does not reveal her presence until the environment – nurture – is pushed to the extreme, be that of starvation or excess.

Sport – in both horses and humans – illustrates the subtlety of the interaction between the two better than almost anything else. It shows how each can become more, or less, important as circumstances change. Often, almost imperceptible distinctions in native talent that are of no importance to the mass of the population make a real difference at the highest level of athletic endeavour.

Francis Galton, the founder of human genetics, himself had an interest in the biology of sporting performance. He studied the inheritance of speed in American trotters (and even developed a method to digitise photographs of thoroughbreds for use by breeders). Galton also looked at the families of oarsmen and wrestlers. The Tyne was then a centre for rowing events and he classified the brawny rustics who wielded the sculls in a series of grades from 'eminently gifted' to those of no particular talent, 'mere conscripts from the race of clubbable men'. Many of the most able came, he found, from long lines of excellent oarsmen. The champion wrestlers of Carlisle and Newcastle, too, tended to belong to families in which that talent was well established. Perhaps, as we would say today, their abilities were in their genes.

Galton persuaded himself that many other attributes ran in families and that they too were under biological control. His Victorian universe had little room for salvation by deeds: for him, original sin marked most of us. In his 1869 book *Hereditary Genius* Galton wrote that: 'As it is easy ... to obtain by careful selection a permanent breed of dogs or horses gifted with peculiar powers of running, or of doing anything else, so it would be quite practicable to produce a highly-gifted race of men by judicious marriages during several consecutive generations ... just as we can breed horses for points, so we can breed men for points.' Genius, oarsmanship, ability and more were fixed at birth and it should be possible to advance our own species into a eugenic future just as effectively as breeders had improved their horses. Each infant, he thought, entered the world with an inborn certificate of merit or its opposite. Galton had no time for Christianity (and got into trouble with the Church about his statistical tests on the efficacy of prayer, which showed no increase in the life expectancy of those regularly prayed for). He hoped instead that eugenics would be 'introduced into the national consciousness as a new religion'.

For horse-racing enthusiasts (if not yet for fans of wrestling, rowing and hereditary genius) breeding is still important. Britain produces five thousand thoroughbreds a year. For many years the model of inheritance used by their owners was absurd. Every starter at Ascot descends from one of three eighteenth-century Arabian stallions of superb quality. Many mares of English origin are in the pedigree, but as they were thought to have no effect on the next generation no attention was paid to their merits. Such chauvinism much hindered attempts to improve performance.

Now, everything has changed. A hobby has become a science. One of horse-breeding's most important tools descends from Galton himself. He had found a close similarity between the height of parents and children and developed statistical methods with which to measure it. Since his day, a mass of information on patterns of resemblance over the generations has been gathered for a vast diversity of attributes, in people, in horses and in many other creatures. Geneticists use it to measure 'heritability', the proportion of total variation within a population that can be ascribed to genetic variation. The figure, which varies from zero to one, can be established for any feature from height to happiness. Human height is very heritable, happiness less so, while the number of offspring tends to score low on the heritability scale.

In flat-racing horses a fifth of the variation in heart rate, muscle power and gait is heritable. The figure is lower over the sticks, while the ability to jump is disconnected from that for speed down the straight. Speed itself gives even lower figures.

Heritability is a much misunderstood measure. First, it deals with variation, and not with absolute value. Any differences in the number of legs in the horse (or the human) population are almost all of environmental origin, for the majority of missing limbs are caused by accidents, by surgery or by external events such as the thalidomide disaster of the 1950s and 1960s. The heritability of leg number is hence almost zero. That does not alter the simple truth that the quantum of legs – be it four or two – is coded for by genes, or that an agent that reduces it even by a single limb will prevent the most well-bred horse or sprinter from ever winning a race. In addition, heritability says nothing about the odds that any particular athlete or steed will succeed for it applies to populations rather than individuals. Most important – and most ignored – the statistic is a ratio, a balance between internal factors and those outside. A change in conditions can as a result much affect its value. In horses, for example, estimates for the heritability of speed made on wet and unpredictable tracks are half those for the same animals on dry and favourable days.

The same rules apply to people. Many of our qualities have

a biological component, but for much of the time environmental noise overrides any innate differences. In Britain, which is among the developed world's most unequal countries, success and failure depend as a result far more on the economic than on the genetic accidents of birth.

Even so, Galton's measure is a useful first step for those who wish to breed horses for speed or cattle for milk. An attribute that always scores low, whatever the circumstances, may be harder to improve by selecting the best as parents than is one with higher heritability. One approach is to keep conditions in the stable or cowshed as constant as possible. With luck, that enables any inborn differences to show themselves. Each top racehorse is so pampered that, in effect, every one experiences the same environment. Inherited variation may then make its presence more obvious and with patient accumulation over the generations the breeders can improve the stock. That may not do much to change the chances of victory on a wet track with its unpredictable risks, but may be more important on a perfect day at the races.

Other sports also provide an insight into the nature, and the limitations, of heritability. Championship athletics of its very nature picks out those at the extremes. As it does, it may reveal biological differences that are in most circumstances overwhelmed by external influences. At schoolboy level, such agencies – sloth, unfitness, fear, ignorance (and in my case an abhorrence of organised enjoyment) – play the largest part. At the top, the rules are different. There, every participant trains to the limit of his or her powers and puts every ounce of mental and physical ability into the struggle. Inborn differences that do not concern most of us then show their effects. Extreme sports such as 'ironman' events (a two mile swim, a hundred mile bike race, followed by a full marathon) cause such stress that success may depend more on the body's innate strengths or weaknesses than it would in a contest limited to a brief dip, a spin on a tricycle and a stroll. As an athlete progresses from school captain to successful amateur, to national title, to Olympic gold, genes become more and more important. Today's ability to identify them might, as a result, change the rules of the game. As it does it will force us to take a wider look at how society copes with inborn differences of perhaps more significant kinds.

Many genes that may alter sporting performance have now revealed themselves. Horse DNA has been read from end to end. Thoroughbreds as a group, long bred for speed, have high-efficiency versions of proteins that influence muscle power, the ability to sense insulin (which controls the rate at which fuel flows to the muscles) and more. That may seem to do no more than prove the obvious – that a thoroughbred will beat a carthorse down the straight – but it also hints at where to look for individual differences between the animals themselves.

Already there are hints that racehorses with particular versions of molecules that help soak up oxygen and burn glucose have a better chance of being first past the post. Other genes work on strength. Half a thoroughbred's body mass is meat. A growth factor called myostatin controls the size of each muscle. Its function is to limit their development as an animal grows or takes exercise. Shifts in its structure lead to changes in physique, for the heavy animals used on farms carry a variant distinct from those of racehorses or ponies. In certain dogs too, a mutation that knocks off a substantial length of the protein has a real effect. A 'bully whippet' is so called because it inherits two copies of a damaged myostatin gene and has muscles twice as big as normal. Most such animals are killed off by breeders. However, whippets with a single copy of 'bully' look much like their fellows and are allowed to survive – and they do far better than average on the track. They are raced in the high-status events of the whippeting world, where almost a third of the runners bear a copy of the variant, compared to just one in fifty of those in the bottom league. Some now say that the mutant animals should be banned because of their unfair advantage.

The most successful thoroughbreds also tend to bear a particular form of that gene and those with two copies are faster, and can maintain their speed for longer, than others not so blessed. The change also influences the animals' favoured distance. Those with two copies of the DNA letter C in a certain position do better in races of less than seven furlongs. Specialists in longer contests, which demand stamina rather than speed, tend to bear the letter T in the same place. The vast majority of horses, together with their relatives the zebras, have two copies of T, which must be the original form. Pedigrees, together with studies of preserved bones, suggest that the speed variant entered the horse population just once, around three hundred years ago, from a British mare. It remained rare until the famous – and fecund – Canadian thoroughbred stallion Nearctic began to spread it through the studbook in the 1950s. That task was carried out with enthusiasm by his even more distinguished son, Northern Dancer (who sired almost a hundred and fifty winners and netted his owners a million dollars each time his services were required). In principle, foals could now be ascribed at birth to the career at which they are best suited and an Irish company called Equinome now offers a test kit that claims to do just that.

Soon, no doubt, many more genes with individually small – but perhaps collectively large – effects on the chances of a win will emerge. A quick check with a cheap gene chip would then add precision to a race-goer's anguished attempts to decide where to take a flutter. A single hair could tell the story. In times to come, prowlers may creep around stables the night before a race to pluck out the evidence. If the punters turn to technology, how can the handicappers, or the bookmakers, avoid it?

Myostatin mutations are found, now and again, in people. A German boy who bears two copies of a damaged version of the gene is unusually muscular (and his mother, who must have a single copy, was once a professional athlete). The gene also contains quite a lot of minor variation. One in twenty Europeans has a certain amino acid in that protein replaced with another. Such people perform rather worse than others when asked to jump up and down for as long as they can, but the effect is small.

Human athletic success involves many talents and, as a result, a great variety of genes. Strength, solid bones and supple joints, the ability to withstand fatigue and pain, and simple obduracy – the triumph of the will – are all essential. Hearts, lungs, arteries, veins and muscles each play a part. At the cellular level, hormones, the nervous system, the rate at which energy is summoned from the reserves and burned are also involved. Many of these attributes have high heritability, as much as 70 per cent when it comes to the amount of skeletal muscle, or the size of the heart. Even the extent to which children choose to take exercise runs in families.

Some rare inherited disorders provide athletic talent as a side-effect. In the 1960s the Finnish skier Eero Mäntyranta gained seven Olympic medals as a result of his extraordinary ability to cover vast distances. He had an unusual form of a gene that controls the numbers of red cells in the blood and, as a result, was more able than his rivals to soak up oxygen. In spite of sneers about fraud his condition was quite natural but, rather like the solitary boy with a variant myostatin molecule, is not very relevant to the population as a whole.

A more general player in the Olympic stakes is called the ACE gene. Its name stands for the initial letters of its protein, the angiotensin converting enzyme. That substance helps to regulate the degree to which blood vessels tighten up or relax when the body is under stress. ACE performs a variety of other tasks. Drugs that inhibit its action are used to treat patients with high blood pressure and are also valuable in certain forms of kidney disease.

The gene is quite variable. The most conspicuous shift involves the presence or absence of a long stretch of DNA that landed long ago in the middle of it. Such changes are of medical importance as those with double copies of the shorter version suffer worse symptoms of heart disease and, if crushed in accidents, have a smaller chance of survival. The shorter version makes up 80 per cent of Africans' ACE genes, rather more than half those of Europeans, but no more than a fifth of those of the people of China.

In a parallel to the horse myostatin story, people with the longer variant tend to do better in endurance sports such as marathons and triathlons, while those with the alternative form excel in pastimes that depend more on short-term strength and power. In that select band of eccentrics who climb the Himalayas without oxygen the long variant is common, so much so that none of the fifteen alpinists who have managed to get above eight thousand metres without a mask has a double copy of the short version, although a quarter of the general European population does. Those with the long form also tend to respond better when they train hard for sports such as the marathon. That version causes the body to generate more heat under hard exercise, which may be why it is less common in the tropics. Perhaps its effects are more beneficial on cold days, or in snowy places. Natural selection has noticed its power. In some Himalayan villages, where oxygen levels are low, the long form is more frequent than in the lowlands.

ACE is just one among twenty or so proteins whose variation has been claimed to influence athletic performance although, as in the case of height or heart disease, there is little consistency among studies and most associations are weak. Some increase the ability to soak up oxygen, while others alter the balance between muscle fibres that react slowly or fast when instructed to contract, with bearers of the former good at marathons and of the latter better at sprints. Many of the variants have no known function, but are more frequent in champions of various kinds. A profile based upon a sample of such genes can, some claim, provide a crude score of overall ability. In a carefully chosen set, just one person in twenty million would have the best possible mix (which means no more than three fortunate Britons) and major players in a variety of sports have been found in some studies to have inherited a favourable combination. This approach is in its infancy but may, one day, help to pick out those primed to reach the heights.

The story of genetics and athletic ability (like that of genetics and everything else) is not simple. Because DNA contains so much diversity some apparent fits with prowess arise by mere chance. Not enough work has been done on non-Europeans (and even ACE itself has little influence on Ethiopian runners). As a useful reminder of the limits of biology in sport, Mo Farah, who won a gold medal for both the five thousand and the ten thousand metre events at the 2012 London Olympics, has an identical twin, Hassan. When they lived in Somalia as children, they would often race each other – and, quite often, Hassan would win. Mo then moved to London to join his father. His athletic potential was realised at once, and he entered a long programme of intensive training which, in the end, allowed him to triumph. His brother remains in Somalia, where he is a successful engineer with little interest in sport.

As is the case for human height, genetic variants may

explain no more than a small part of the total variation in ability to run, to jump, or to throw metal balls. Even so, Olympic races are often won or lost by fractions of a second and there, minor as their influence may be under most circumstances, genes may be of real importance.

The notion of an inborn destiny in the race through life extends well beyond the athletics track. In 1955, at the age of eleven, at Grove Street Primary School, in the Merseyside suburb of New Ferry, the quality of my own DNA was tested. Those who administered the test did not realise they were doing genetics (and the structure of the double helix had been announced just two years earlier) but the people who designed it were true descendants of Francis Galton.

The 1944 Education Act introduced the eleven-plus examination. It was based on the logic of *Hereditary Genius*; that Britain contained a hidden pool of native talent that should be drawn upon to pick winners who could be trained to the limits of their ability, together with losers, upon whom the state needed to exert less of an effort. The Act set up three kinds of school, secondary technical, secondary modern, and grammar (although in practice just the last two emerged in any number). For children still a couple of years from puberty the questions were quite tough: 'In each of the sets of words given below there is one word that means something rather different from the other three. Find the different word in each line and write it down: 'scarlet, blue, red, pink', 'sewing, cotton, needle, calico', 'firm, rough, solid, hard'. Another task was to give an account of an imaginary conversation between an eagle and an owl.

Such demands were perhaps better understood by a middle-class child such as myself than the average inhabitant of proletarian New Ferry, who was less likely to be familiar with needlework or the habits of owls. That may be why I was among the half-dozen students, out of a class of thirty or so, who gained a place in the local grammar school. Across England and Wales, around a quarter of state primary school children got through the test, but the proportions varied from fewer than 10 per cent in parts of the North-East to two in three in Merionethshire.

The eleven-plus was claimed, with almost no evidence, to identify children blessed with inborn talent. Had a molecular test to search them out been available the authorities would no doubt have used it. The examination was not meant to be passed or failed, but instead to determine for all pupils the education best 'suited to their abilities and aptitudes'. Parents saw through that straight away. For their children, the day they sat the paper was the most important in their lives, for (unlike racehorses) it was not the talented but the supposedly less so who were handicapped for life. I became an academic but my brother failed the exam, went to a secondary modern, and spent his career as a bricklayer.

Education in those days was a marathon in which fewer than 5 per cent of pupils in the state system made it to the finishing post, a university degree. For those who stumbled at the first hurdle, progress was almost impossible. When, seven years after my ordeal of colour, cloth and avian life I did my Advanced levels (the key to university entrance), no more than one in a thousand of the students who had failed the eleven-plus in 1955 even sat those papers and not a single such child among their hundreds of thousands got into Oxford or Cambridge. Perhaps, as the testers assumed, they lacked the native wit so to do; but perhaps the huge differences in quality within British education also played a part. Since then there has been an attempt to build a less polarised system, with comprehensive schools which most local children attend. A few authorities have kept grammar schools, but the overall performance of their region's pupils is worse than in places that have moved to a more equitable scheme. Almost none of the countries that rate high in international educational tables practise selection, but successful as comprehensives have been, the pressure is on once more to return to inequality.

For some lucky pupils there was, and still is, a convenient escape from the rigours of fair competition. Even the many parents convinced of the innate superiority of their own progeny are happy to improve their educational environment if they can afford to do so. The (notably level) playing fields of Eton have produced great scientists, authors and sportsmen, as well as their large constituency of fops, fools and fraudsters. Almost without exception, they are the scions of the rich, who go to great lengths to ensure that their sons inherit a sense of entitlement as well as the cash, cachet and cut-glass accent needed to indulge in it.

The bracing effects of a canter around an expensively smoothed pedagogical track can be spectacular. In 2010, just five top British schools (four of them, Eton included, private; the other an exceptional state selective sixth-form college in

Cambridge) sent as many pupils to Oxbridge as did two thousand state-funded comprehensive schools and colleges combined. The power of their privilege extends to sport (as well as, needless to say, to business, to journalism and to politics). More than half the British gold medallists at the 2008 Olympics attended private schools and in the 2012 Games the fraction was not much lower. In cricket, two thirds of the England team had such an education, a proportion up from just one in ten a generation ago. Either the private establishments are remarkably good at spotting native talent, or they offer an educational environment far better than that available to plebeians. Their annual fees of thirty thousand pounds or so suggests that an expensive dose of nurture does play some part in enabling whatever capacities are perched upon their pupils' pinstriped backsides to manifest themselves.

Wealth itself is, of course, highly heritable; and the more unequal a society, the more it stays within a family line. The Gini Index is a measure of a nation's divergence in income between top and bottom. At zero, everyone is the same and at 100, the richest person has all the money. According to the Organisation for Economic Cooperation and Development, Denmark scores 23, while Britain and the United States have benefited from decades of political progress and click in at 34 and 38 respectively (and for each the figure is rising fast). The centres of inequity are in South America, with some countries with a score of over 60, and Africa, which does not do much better. The tie of equality with national educational level is clear. Across the globe there exists a good fit, independent of the association with each country's actual wealth, between the Gini score and that of average intelligence.

The poor pay a heavy price. In the year 2000 the British government began the Millennium Childhood Study. It will sample a cohort of children throughout their lives. By the age of seven, the average scores of deprived children were already lower than those from richer families for mathematics, for comprehension and for reading ability. They received less attention from their parents and, on average, mothers were younger and their babies weighed less at birth (itself a predictor of reduced mental ability). They were less likely to be breast-fed and, quite often, the father was absent. Poverty reduces ability for reasons that have nothing to do with DNA.

Whatever the power of privilege, no biologist denies that genes are involved in differences in intelligence. Research on the subject was once the haunt of the obsessed. It was filled with accusations of dishonesty, bias and racism from both sides. Now it has become a respectable part of psychology. IQ score has a heritability that hovers at around 50 per cent (even if many who point at the figure appear not to know quite what it means). Like height, the measure is malleable, with an increase at around three points a decade over almost a century. A typical modern boy transported back to the outbreak of the First World War would have an IQ of 130, while an average soldier on the Western Front, moved forward to today, would score 70. Many of those who once used the idea of 'intelligence genes' to demand that less be spent on the irredeemably stupid (by which they usually meant blacks) have failed to notice this.

Schools have without doubt played a part in such improvements. Almost everywhere, in the past few decades, the number of years in the classroom has increased and the score has risen to match. In the 1960s Norway raised the age when pupils left school from fourteen to sixteen. When the two cohorts were tested at the age of nineteen, the younger group had gained four points for each extra year of education. The new world of television and of electronic games and the instant availability of information may also have made young people better at the tests.

All this shows how practice improves intelligence, as it does athletic performance. The global change with time – and the great differences in average levels that still exist across among countries – can also be tied to a shift towards better physical conditions. IQ score responds to good food in infancy in much the same way as does height. There is, as a result, a correlation between individual or national stature, and average mark. It also shows a fit with infectious disease, even when the effects of wealth and of education are removed. The brain is expensive. A newborn uses nine-tenths of its energy on it and a child who struggles with infection must also summon up energy to fight it off, while illnesses such as dysentery stop the infant from absorbing a meal in the first place. The control of parasites and of hunger may explain part of the global increase in intellectual ability over the past century and some of the continuing differences between rich and poor nations.

The environment – wealth most of all – also makes a great difference to the genetic contribution to IQ. A survey of American twins, some raised in real hardship, others in middle-class homes, and yet more the children of the affluent shows that the importance of the double helix differs greatly between the poorest families and the others. In the poverty-stricken households, heritability was almost zero, no doubt because a desperate lack of proper food, affection and culture drove everyone down to the same level. In rich or bourgeois families the figure was much higher. Just as in sink secondary moderns rather than English private schools, or wet rather than dry race-tracks, poor conditions overwhelm genetic potential.

When a working-class child is adopted into a bourgeois family his or her test score rises by, on average, fifteen points. Sometimes, several unrelated infants join the same household. As they do not share genes with each other, or with the children of the parents who adopt them, any similarities above the random level must reflect a common environment. In fact the adopted and the natural children come to resemble each other closely for IQ score when they are small, but after adolescence the similarity disappears. The heritability of that measure rises to as much as 0.7 in people over retirement age. In other words, the test results of young children who live with books and educated parents increase in response to their favourable circumstances, while those of youngsters adopted into poor and indifferent families do the opposite. Later in life genes make their presence more obvious.

The eleven-plus examination hence did an excellent job of picking out the already privileged but failed entirely to identify deprived infants with a modicum of untapped talent. Half a century after I walked into the examination hall, thousands of children still face the same ineffective hurdle at much the same age, with much the same pernicious results.

If every Briton went to Eton (or for that matter to Grove Street Primary) individual variation in intellectual ability would be determined much more by genes than it is today. The double helix would also have more influence if the environmental playing field in terms of parasites, food, or literacy were levelled. There might then be, as the Education Act of 1944 proposed, an objective test of native potential. At present, there seems no hope of that and the British educational system is, in its chaotic and unfair way, set to continue to preserve privilege (and its opposite) for future generations.

Individual IQ, whatever lies behind it, predicts many things. Those with low scores tend to be poorer and less happy than average, to die younger and to become more involved in crime. Francis Galton was not slow to notice. As he wrote: 'The ideal criminal has marked peculiarities of character: his conscience is almost deficient, his instincts are vicious, his power of self-control is very weak, and he usually detests continuous labour. The absence of self-control is due to ungovernable temper, to passion, or to mere imbecility, and the conditions that determine the particular descriptions of crime are the character of the instincts and of the temptation. The perpetuation of the criminal class by heredity is a question difficult to grapple with on many accounts ... It is, however, easy to show that the criminal nature tends to be inherited.'

The idea that delinquency is linked to native imbecility,

passion and temper remains popular. In some places the law has acted as if that were an established truth. As late as 1976, Sweden sterilised prisoners as a condition of release. In Japan a 1948 act meant that a 'genetic predisposition to commit crime' was grounds for castration. Many other countries have had such policies, but they have faded away.

Nobody denies that crime is heritable. In the UK the strongest predictor of whether a boy will end up in prison is whether his father has done time. Even dedicated hereditarians accept that social pressures are in part to blame, but so strong is the disposition to search for a biological cause that plenty claim to have found genes that increase the likelihood of offence. They may even, in part, be right.

As in the case of human height or intelligence, scans of the whole genome in a search for variants that might be more frequent among criminals as a class have produced weak and inconsistent results. Even so, a few people at the extreme end of the distribution (as is the case for dwarfism, gigantism and cross-country skiing) have errors in particular genes that may much change their fate.

Many of those who end up behind bars have a psychopathic personality. They lose their temper at the slightest provocation, have no concern with the distress of others and care nothing for the pain they cause. A score on the psychopathy scale is about as heritable as that for IQ. Such people are twenty times more frequent in prisons than they are in the general population and a few may have faults in particular stretches of DNA.

The monoamine A oxidase (or MAO-A) gene makes an enzyme involved in the transmission of nerve impulses. Some members of a Dutch family have an inactive form of that protein, and several among them have been in serious trouble with the law, for they fly off the handle at the least provocation. They show no interest in the fate of their victims and no regret for their actions. The press hailed the discovery as 'the gene for crime', but that severe error has never been found in other psychopaths or in anyone else.

Even so – in much the same way as myostatin alters athletic performance – normal variation in the same gene may, under some circumstances, have an influence. The MAO-A gene shows individual differences both in structure and in activity. Among its tasks is the regulation of serotonin, a nerve transmitter that influences mood, appetite and alertness. In crude terms, high levels of that substance lead to contentment and low levels to despair. In a survey of young New Zealanders those with reduced activity of the MAO-A enzyme were more likely than average to get into trouble with the law, and many among them showed little concern for the misery they caused. Again as in sport or in schooling, in spite of that genetic predisposition the environment was much involved, for most people born with low activity lived exemplary lives. Only children exposed to stress in their formative years – poverty, physical or sexual abuse, or expulsion from school – got into trouble. Any child may run into difficulties when faced with such problems but while the rate of offending went up by three times in those with abusive lives but normal monoamine oxidase, it rose by almost twenty times in troubled children who inherit low levels of enzyme activity. Once again, nature and nurture work together to make us what we are.

Levels of crime in affluent New Zealand are at or below those of most European countries. There and across the globe the patterns of violence and theft, like those of IQ score, height and Olympic medal count, show a close fit with national levels of inequality, of hunger and of disease. In parts of Central America drugs, poverty and political instability mean that the murder rate is sixty times that of Europe, at eighty per hundred thousand people (which means that 2 per cent of men aged between twenty and thirty will be murdered during that risky decade of their lives). In the United States the overall figure is around five per hundred thousand; in Britain between one and two. Singapore suffers just one murder per three hundred thousand each year, and is two hundred times safer than El Salvador.

Guns, inequality and despair are no doubt important in all this, and variation in enzymes such as monoamine oxidase might even play a small part. One gene's effect on violent crime (and on much more) is so obvious, and so large, that its power is ignored almost by default. Mendel's Elephant, as the culprit might be called, has been in the room for so long that its presence is scarcely noticed. It bears an important message for those who must cope with the interactions of genes with society.

The elephant is male. From Singapore to Central America, whatever the actual incidence of murder, those who bear a Y chromosome kill and carry out other crimes at ten times the rate of those who do not. 95 per cent of British prisoners are men, as are nine out of ten of the nation's psychopaths. Every claimed biological disposition towards crime is more powerful among men than among women. In the New Zealand study, for example, the joint effects of stress and low monoamine oxidase activity manifested themselves in boys but not in girls.

The genetical pachyderm is at home in many other places. In the Sport of Kings, fillies, colts and geldings (castrated males) often compete in separate events to ensure that the results are not too predictable, and in mixed-sex races handicapping is used to give the females a chance. Men outperform women in most sports. At Olympic level, from the hundred metres to the fifteen hundred metre race, men are about 10 per cent faster than members of the opposite sex. In the high jump, long jump and pole-vault the male advantage is even greater, with male records at least a sixth higher than those of females (my own nemesis, the shot-put, cannot be compared as males are forced to wield balls half as heavy again as those hurled by females). In the marathon, often a mixed event, male champions come in some ten to fifteen minutes before the first women (even if the fittest females beat a large constituency of men into the ground). Without doubt, much of this contrast comes from the manifold effects of testosterone on body and mind (a fact manifest in the severe penalties applied to those who abuse that drug to improve their performance).

On IQ tests, too, males and females differ, but the effect is hard to assess because they each do better on distinct parts of the questionnaire. Men excel on numerical questions or when asked to rotate a complex shape while women do better on more literary elements. The balance of the two is adjusted in an attempt to even the scores. Even so, for many years women did rather worse than their male contemporaries. In 2012, for reasons unknown, that pattern was reversed.

The law, too, takes a great interest in the Y chromosome. It adjusts its responses to match. The 2000 Equality Act demands that all public organisations (courts included) should promote equality of opportunity of men and women. The policy has not been a success. Males and females do offend at different rates and are involved in distinct kinds of crime, but even when that is allowed for the penalties are far from equal. Men tend to be blamed, rather than forgiven, for their biology. When like is compared with like, twice as many women as men receive a discharge after a criminal offence and many more of the former are given a community sentence rather than custody. Female shoplifters and drug dealers are less likely to be imprisoned than are their male counterparts and women who commit violent offences face just half the chance of immediate jail as do their masculine equivalents. The tendency for females to receive shorter sentences goes back to Victorian times. Many of those who can recall the days of the death penalty, now – thank God – no longer with us, remember the name of Ruth Ellis, in 1955 the last woman to be hanged in Britain, but who has any recollection of Peter Allen and Gwynne Evans, who nine years later were the last men to suffer that fate?

Fewer female offenders have a criminal record and they tend to commit less heinous crimes, but even when this is corrected for the difference remains. Magistrates often justify their decisions on the grounds that the unfortunates in the dock are more troubled than troubling, or express concerns about the effects of imprisonment on their children. Women, say those on the Bench, are more respectful in court, and this too counts in their favour. Even so, and whatever the rationale, those without a Y are deemed to be less culpable for their acts and individuals disadvantaged by bearing that potent structure as more so. In the end, all those criminals, male and female, are being judged on genetic grounds.

Biology has helped to bring sport, education and the law up to date, but the biological and theological difficulties associated with inborn physical and moral weaknesses remain. How can there be equality when some are given privileges, or face handicaps, because their DNA predestines them to win or lose? Society is less than consistent. Male criminals are penalised for their birth-right but a child blessed with inborn intellectual gifts is given better treatment than his less fortunate fellows. On the sports field too, attitudes to handicapping are mixed indeed. Scientists now understand some of the biology that lies behind the variation in human talents but society still holds to prejudices that reach back to before science began. Unfortunately, Pelagius, St Augustine and John Calvin are no longer around to advise us.

CHAPTER THREE

THE BATTLE OF THE SEXES

William Blake, *Visions of the Daughters of Albion*

*Behold, a virgin shall be with child, and shall bring
forth a son.*

GOSPEL ACCORDING TO ST MATTHEW 1:23

I am a clone, the son of a clone, and like many of those who
share that distinction, an evolutionary dead end.

My own body cells copy themselves without benefit of sex,
and of all the untold billions of sperm my feeble frame has
produced since adolescence, not one has found an egg with

which to fuse and to generate some reorganised version of myself. My mother, as it happens, was an identical twin (although I do not think that is why I became a geneticist) and even if I did sometimes, as a child, find it hard to tell her and her sister apart, their uncanny similarity never struck me as raising a theological issue. For some people, it does; and has done so since Eve was cloned from the rib of Adam.

It has to do with the soul. According to the American physician Duncan MacDougall, that essential organ weighs just twenty-one grams. In 1901 he put the beds of several aged patients on a sensitive balance and measured the loss of *avoirdupois* at the instant of death; the moment when, he was sure, the spirit departs the body (he did a control experiment with dogs, which showed no such lightening on their demise). As he wrote, 'It is unthinkable that personality and consciousness can be attributes of that which does not occupy space.' He had proved, to his own satisfaction at least, that the soul had a physical existence.

If it does – or even if it does not – how does the magical essence get into the body in the first place? When, in other words, does life begin? That question has exercised thinkers since Old Testament times and still impinges on issues such as abortion and stem cell research. In Exodus, death is prescribed for a man who strikes a woman so hard that her foetus dies, as an indication that the unborn child was seen as human (if the infant survives the malefactor is punished, but allowed to live). Jeremiah, too, quotes the Lord's statement that 'Before I formed thee in the belly I knew thee; and before thou camest forth out of the womb I sanctified thee, and I ordained thee a prophet unto the nations' as further evidence that the fertilised egg has a soul.

To the medieval theologian St Thomas Aquinas, in contrast, true life had to wait. His claim, restated in modern terms, was that the essence of humanity is under the control of the Y chromosome, for he argued that boy foetuses gain their mystical structure at forty days of gestation, while females have to wait more than twice as long. The emergence of an infant was a gradual, rather than an instantaneous, process: '... the soul is united to the body as its form. Therefore, the soul does not exist before ... the organisation of the body.'

Today's Catholics reject that argument. Soul and sex go together from the very beginning and the essence of humanity is inserted at the instant of fertilisation. In the words of Pope John Paul II, 'The Church has always taught and continues to teach that the result of human procreation from the first moment of its existence must be guaranteed that unconditional respect which is morally due to the human being.' His doctrine encouraged the anti-abortion movement, but also caused my late mother's status to become ambiguous. Identical twins arise when an embryo splits well after sperm met egg — but if that is the case, where does the second soul come from? Were my mother and her sister, my Aunt Pegi, blessed with just half a copy each, or does God have a stock of spares ready to insert when needed? In a mirror-image of that dilemma, many people are chimeras; they contain cells from two individuals, non-identical twins, whose embryos fused together in the first few days of development. Do they gain an extra helping of the divine afflatus as a result? In rare cases, a foetus receives almost all its genes from the father alone. Does it have just half the normal dose?

Those questions point at the biological, as much as the theological, problems posed by sex. That habit began – as Genesis confirms – not long after life itself. Why use such an expensive mechanism given that cloning, through twins or otherwise, is, at least from a woman's point of view, more effective? And if two or more infants from a single cell are possible, why not take a step further and abandon males altogether? Sex is filled with conflict; between males and females, within each gender, and between parents and children. Why bother with it? The Book of Proverbs is baffled: 'There be three things which are too wonderful for me, yea, four which I know not: the way of an eagle in the air; the way of a serpent upon a rock; the way of a ship in the midst of the sea; and the way of a man with a maid.' Biology shares that final sentiment.

The cost of sex involves much more than the effort and annoyance associated with the act itself and the genetic events behind it. Liaisons with males force females to squander their energies in copying the genes of another individual, and to dilute their own investment with progeny who carry his DNA. Almost as bad, any female with a happy combination of genes is unable to hand them on unaltered to the next generation but instead must reshuffle them with those of a partner to produce a random mix. To allow males into the equation also enables the spread of parasites (and not just sexual disease but parasites of the genes). Around half the genome consists of such intruders, which do little and cost a lot to copy.

The ways of men and maids still present some of biology's most intractable problems. The question of how sex began is not the same as that of how it is maintained, of why there are no more than two players on the field rather than dozens, of why males and females exist in equal numbers, or even of why they often look so different. We have partial answers to some of those queries, but the issue of how the business got off the ground in the first place remains unresolved. At least twenty theories are in circulation, none altogether satisfactory.

The priesthood is just as confused. The Song of Solomon uses overt sexual imagery as a parable of the relationship between God and Man (Joseph Smith, founder of Mormonism, was so shocked by its eroticism, with 'lie all night betwixt my breasts ... thy navel is like a round goblet ... breasts are like towers', that he missed the whole Book out of his 'Inspired Version' of the Bible). On the other hand, in most of the Good Book's verses the act of generation is hedged around with euphemism and prescription. Every creed has its own rules. Some branches of Islam insist that all children should be the progeny of a known mother and a known father which means that anonymous sperm donation is banned. Judaism became more magnanimous with time, for the era of ritual stoning of adulterers (rape victims were spared if they could be proved to have resisted) has gone. Now, unless a particular practice is outlawed in the ancient scrolls it is – from in-vitro fertilisation to surrogate motherhood – allowed.

Christianity tends to the dogmatic in the bridal chamber, with Catholics more censorious than most, while the eighteenth-century Shakers, an offshoot of the Quakers, forbade sexual intercourse altogether (at their peak they numbered thousands but now just three are left). Buddhism does not allow its monks to indulge and recommends that all its followers behave with appropriate restraint.

There is plenty of love in the Bible, but in its sexual context the emotion is almost confined to the Old Testament. In the New, it is directed instead towards Christ, or God, or at fellow citizens, with few hints of its biological function. St Paul describes marriage as a *sacramentum magnum* – a wonderful mystery – but in his Epistle to the Galatians a quarter of his paths to Hell were paved with sexual intentions: 'Now the works of the flesh are manifest which are these: adultery, fornication, uncleanness, lasciviousness, idolatry, witchcraft, hatred, variance, emulations, wrath, strife, seditions, heresies, envyings, murders, drunkenness, revellings, and such like: of the which I tell you before, as I have also told you in time past, that they which do such things shall not inherit the kingdom of God.' In his missive to the Corinthians he was so convinced that the Second Coming was near that he felt that even those who were married should give up the habit: 'But this I say, brethren, the time is short: it remaineth, that both they that have wives be as though they had none.' He had just one piece of good news: 'If they cannot contain, let them marry: for it is better to marry than to burn.'

His descendants were just as censorious. St Ambrose, the fourth-century Bishop of Milan, wrote that 'Happy are those who are virgins', while St Augustine, who had himself left his mistress and children in search of a purer life, felt that continence was preferable to marriage, but marriage was better than fornication. Others felt that humankind was doomed because, in order to reach true sanctity, intercourse would have to be abandoned altogether. That censorious view stands in stark contrast to the Judaic celebration of sex as a means of increasing the numbers of the Chosen People.

The Catholic Church still has a convoluted connection with the marital bed, and with what it accepts as 'natural' (and hence to be allowed). The advances of medicine mean that it has been dragged into some unexpected conflicts. Rome is against abortion and human cloning but it also disapproves of in-vitro fertilisation; 'a technology that wants to substitute true paternity and maternity and therefore that does harm to the dignity of parents and children alike' (nevertheless, five million infants have been born with its help). In an attempt to avoid the moral pitfalls of the test tube, the Church recommends GIFT – gametes inserted into the fallopian tube – instead. The sperm is collected not in the obvious way, but as a result of ejaculation into a condom (with a hole in it to give a few of the agile cells a chance to do their job in the approved manner) at the time of intercourse. The egg is mixed with that useful fluid and replaced at once in the female reproductive tract so that fertilisation can take place where God planned it. The technique has not proved popular.

Several of the Apostles were married (and a – much-disputed – early scroll hints that Jesus himself was the husband of Mary Magdalene) but chastity was soon adopted as one of the Seven Virtues of the Catholic Church. Clerical celibacy took hold from the fourth century and the Vatican enjoins its acolytes to avoid sex 'for the sake of the Kingdom of Heaven'.

Millions of men and women have mortified their flesh through abstinence (even if some female celibates have a 'marriage with Christ' as a consolation prize).

The Old Testament is much concerned with virginity, whose importance is constantly emphasised. As is the case for so many of its precepts, its concern with purity leaks over into its successor, the New. Did, for example, the Virgin Mary with her miraculous infant herself arise without sin, through an immaculate conception? Aquinas felt that she was besmirched by her parents' desires but others insisted that she was without flaw from the start.

In 1854 Pope Pius IX issued a Bull that lifted the taint of copulation from the holy child's mother: 'The doctrine which asserts that the Blessed Virgin Mary, from the first moment of her conception, by a singular grace and privilege of almighty God, and in view of the merits of Jesus Christ, Saviour of the human race, was preserved free from every stain of original sin is a doctrine revealed by God and, for this reason, must be firmly and constantly believed by all the faithful.' In other words, the Virgin, through her son, was absolved from the error of Adam and Eve, and preserved immaculate.

The notion that gods are too pure to become involved in the messy business of sex is widespread. In the temporal realm, such an origin has been claimed for several pharaohs, for Greek emperors, and even for Alexander the Great. However, some Christians are not convinced even of the asexual origin of Jesus himself. St Paul adds a certain ambiguity when he writes that the Saviour was born 'from the seed of David according to the flesh'. Thomas Jefferson, devout as he was, wrote in 1823 that 'The day will come when the mystical generation of Jesus by the Supreme Being as his father, in the womb of a virgin, will be classed with the fable of the generation of Minerva in the brain of Jupiter.'

In 1899 an American embryologist discovered that he could persuade unfertilised sea-urchin eggs to develop if he added salt to the liquid in which they were held. The *Boston Globe* reported the event as 'Creation of Life ... Lower Animals Produced by Chemical Means. Process May Apply to Human Species. Immaculate Conception Explained.' He received offers from women to donate their eggs, to 'free them from the shameful bondage of needing a man to become a mother'. His work became the foundation of modern developmental biology. That field made so much progress that it has become possible to grow individual cells from a fertilised egg and to use them to replace damaged adult tissues.

In fact, virgin birth is common even without salty water. As the old joke has it, in Genesis God says to life 'Be fruitful and multiply' but most of the time it divided. In sexual creatures, at fertilisation two cells unite to make one, but at all other times a single cell splits to make two. Everyone is a cloned copy of the fertilised egg that made them and almost every one of their thousands of millions of cells is an identical version of their original, changed only by the slow accumulation of error. Sex is cloning in reverse, for each party to the transaction passes on just half their genes, which are scrambled into new combinations.

Much of the Old Testament is the tale of a people obsessed

by the act of generation. Both its heroes and its villains indulge in practices that would seem repellent nowadays. They include polygamy, incest, prostitution of daughters by their fathers and wives by their husbands, and mass murder of the virgins of defeated tribes. The Phoenician princess Jezebel, who helped to lure her husband away from the true path to grace, was accused of seduction ('she painted her face, and tired her head') and thrown to the dogs. Some of the era's marital arrangements would raise more than an eyebrow in modern times. The daughters of Lot slept with their father when he was drunk, and Abraham married his half-sister Sara while Job entered cousin marriage twice, with two sisters.

In a backlash against such habits, Leviticus and Deuteronomy set out a series of sexual prohibitions. Some have to do with the dangers of contact with menstruating women or with men with damaged testicles, but others pronounce the death penalty for a husband who sleeps with both a woman and her mother, as well as for those who 'lie with mankind as with womankind'. It also dictates who can copulate with whom: 'None of you shall approach to any that is near of kin to him, to uncover their nakedness.' The prohibitions of the unclothed included aunts, uncles, and daughters-in-law. Christianity had an even greater detestation of such acts, and for a time construed the rules with such rigidity that even marriages between fifth cousins were forbidden. After the Reformation, such bans were lifted in most places. Now, in only a few societies, such as China, North Korea and half the fifty United States is cousin marriage forbidden by law (Texas introduced its legislation in 2005).

Sex and strife have been bed-mates since the Serpent made its promise. As soon as it was accepted the process of reproduction was seen as polluted. No longer naked and unashamed, Adam and his partners' eyes were opened to their disgrace: '... and they knew that they were naked; and they sewed fig leaves together, and made themselves aprons.' Genetics has discovered when embarrassment was invented. The desire to shroud ourselves began, as Leviticus might have predicted, with uncleanliness.

People, like chimpanzees, have lice. Our own come in two distinct forms, the first specialised to live on hair and the other on clothes. The build-up of mutations between the chimpanzee and the human louse since the split of the primate lineages around eight million years ago can be used as a timer to estimate when the distinct forms of the human parasite diverged. That must have happened soon after we donned our first garments. The molecular clock shows that the two versions made the break around a hundred and twenty thousand years before the present, at about the moment that truly modern peoples first appear. That was an era of social change, of a new culture of stone tools and, perhaps, of the first attempts at language. Shame about sex, and the emergence of 'aprons', may have been part of the package that made us what we are.

Genesis tells us that the habit began on the third and fifth days, when plants and then animals first brought forth others after their kind. Since then it has caused endless trouble. Both science and myth see sex as full of conflict, compromise and contradiction. The process must balance the interests of males and females, parents and children, and instant gratification versus delayed reward. Its origin, its rules, and the circumstances under which it should be exercised form a large part of both the biblical and the biological narrative.

Parts of the story are clearer than they were. Why, for example, are there just two sexes? It seems rather inefficient, for if there were dozens, or hundreds, almost everybody would be available for intercourse rather than, as at present, the choice being limited to a single tedious variant on one's own gender. The answer goes back to the first days of the cell, with the transformation of aliens – bacteria – into mitochondria, those tiny power-stations. Mitochondria are passed on by females alone, and the ability to do so is almost a definition of what femininity means. The relationship of the two players started almost as parasite and host and remains finely tuned. Any incursion of mitochondria from a third party would be fatal, for in the fertilised egg their genes would compete for attention with those already in residence. As a result, just one segment of the community – females rather than males or members of some hypothetical third, fourth or fifth sex – hands them on.

Both sexes are under pressure to make as many reproductive cells as is possible. The egg must transmit other cell structures, together with a reserve of food for the first hours of development. That means that it cannot shrink below a certain size. Males, in contrast, are under pressure to make their sperm as small, and as numerous, as they can in the hope of meeting a cell of the opposite sex; and any version that does the job better than others will take over. Once again, a third player would get no chance to squeeze in. The masculine race to the bottom and feminine attempt to stay afloat meant that the descendants of Adam evolved to be active and aggressive, and those of Eve to be coy and careful. That ancient dichotomy still influences men and women today.

And why are there so many sperm-donors? Each time a husband sleeps with his wife he makes enough of those cells to fertilise every female in Europe. Why not employ a busy bureaucrat in a Brussels bordello instead? It has, once again, to do with the divergent tactics of the two players. If, in the Garden of Eden, the Lord had taken all Adam's ribs and had fashioned twenty-four Eves rather than one, the First Man would have been happy indeed, for he could have affairs with a multitude of partners. The Eves would not be so well off. Even if they attracted the lone male's attentions they could, for simple reasons of biology, have not more than about a dozen children each while Adam, with his multiple spouses, might have had hundreds. As a result, if males are rare and females common, genes that produce maleness are favoured. Should the numbers of men rise above that of women, the balance is reversed.

At birth there are, in Britain, a hundred and five boys to every hundred girls and just after puberty the figures are equal. Males continue to shuffle off the mortal coil faster and faster until, at the age of eighty, the ratio stands at two women to every man. In parts of India and China, in contrast, the preference for sons is so ingrained that, because of abortion and the murder of girl babies, the ratio of males to females at birth is as much as a hundred and forty to one. As a result, within a decade or so those nations will have a surplus of tens of millions of young men, many of whom will fail to find a partner. Already forced marriage and rape are rife, proof that the reproductive value of females rises when they are rare.

That divergence of interest is just part of a vast arena of disagreement. The struggles go on before, during and after the sexual act itself, and persist throughout pregnancy, after birth, and even unto the next generation. Molecular biology has made sexual reproduction into an even less seductive pastime than it appeared to the Church Fathers.

Most of the beauties of nature — birdsong, blooms, baboons' bottoms — are cues of masculine appeal. They are expensive to make and females also pay a price as they avoid the attentions of insistent but inferior partners.

Such delights are but an overture to an opera of conflict. Post-copulatory disagreements are often more bitter than are those on the stairs to the bridal suite. After sex, a male may guard the female against the attentions of a second suitor. His techniques can be crude. Plenty of creatures (cats included) become physically attached to their mate for a period; a position ridiculous but effective, for it ensures that their own sperm have a head start. A certain Australian beetle does the same as it hangs single-mindedly on to its female until it can be confident that she has been fertilised. It has the same response to the shiny beer bottles that litter that nation's highways.

Other male insects produce poisons with their sperm that force the female to invest more in her brood than she might wish. Yet others make a great excess of the vital fluid to flood out the contributions of a predecessor, or have an artfully crafted penis that extracts his donation. In defence, the recipient may divert sperm into a blind alley and choose which lucky swain she allows to fertilise her eggs. Certain snails go further, for their male fires substantial darts coated with hormones into his partner to force her to accept his own sperm at the expense of that of any rival. In some mammals sperm itself contains a hormone that persuades a female to ovulate.

Humans have calmer reproductive lives than do snails and are restrained even in comparison to most of their relatives. In the Western world, DNA tests show that fewer than around one birth in forty or so is the supposed father not the child's real progenitor. Men have smaller testes and lower sperm count than many other primates, some of which have frenzied copulations of a single female with many males. A male macaque was once observed to mate forty times in a day (which was more than the champion stud among the thousands of men interviewed by Alfred Kinsey for his book *Sexual Behavior in the Human Male* managed in a week). Human behaviour is closer to that of the gorilla, in which the dominant male enjoys the favours of several females but most females mate just once (a pattern that leads to vicious battles among the males themselves). Copulation takes less time than in the orang-utan, while the process of sperm manufacture is slower than in almost any other mammal.

Many have claimed to find conflicts between human sperm cells. Their results look less attractive in the cold light of scientific dawn than they do through the rosy spectacles of late-night sociological speculation. The once popular idea that 'kamikaze sperm' attack another's cells within the female reproductive tract is based on wishful thinking. There is also no evidence that men adopt the tactics of certain small mammals, in which gangs of sperm unite into a scrum that bullies its way forwards at the expense of those who strive in solitary splendour.

Some assert that men who have been away from their partner flood her with more of the vital fluid than those who are in regular contact with her and can as a result be fairly confident that she has not strayed. Sperm might also, it has been claimed, act as a sedative to ensure that a female lies back in a light doze after the crucial moment to increase the chances of fertilisation, and as an anti-depressant to titillate her interest in a later bout. Semen coagulates after delivery, perhaps (as in many of our relatives) to form a plug to deny entry to the contributions of a second male.

When it comes to the biology of sex, erotic fantasy is never far away. In terms of penis size we are nothing special – but why the odd shape? Meticulous experiments with model organs and with fluids of different degrees of stickiness hint that it could act as a reverse-action pump to scoop out any foreign semen that might be present before a new dose is inserted (the prepuce has, the researchers suggest, evolved as a sneaky trick that enables a male to have his own sperm carried onwards to a second female by a later rival). This theory does not hold water, for the organs of many primates, monogamous and promiscuous alike, have more or less the same profile as our own.

In a series of subtle negotiations, women fight back against the wiles of men. Their partners are forced to struggle to keep up with them in a contest that takes place before, during, and long after the act of sex itself.

Human females, unlike most of their primate relatives, do not tell the truth about when they are fertile. Female chimpanzees flaunt swollen backsides and genitals for the several days in each cycle when an egg is ready to be fertilised. That elegant display sparks off intense competition among their swains, who attack each other and in the intervals groom the potential partner. Such signals might increase her chances of success with the best possible mate after the rivals have fought themselves to a standstill or might ensure that many males want to mate with a female so that none can be certain who the real father might be. A male will then avoid killing off a brood in order to put its mother back into the copulatory arena.

A hint of the origin of woman's coyness comes from another twig on the primate tree. The pygmy chimpanzee has a more relaxed life than does its closest relative. Its males fight less, have lower levels of testosterone, and are smaller with less fearsome canines. Unlike chimps proper, the females solicit sex with a partner. Their parts do swell around the time of fertility, but the fit is less precise than in their close relative and, quite often, males who mate outside the advertised period succeed in their efforts. The pygmy chimpanzee female has begun to hide the truth about her availability and has gained a certain liberation as a result.

A woman produces an egg about two weeks after her last

menstruation. As sperm can survive for seven days in the vaginal canal, that gives a week or so in which sterility is more or less guaranteed. The most fertile phase starts soon afterwards, and lasts for about another week. The Catholic Church accepts that married couples can use this information as a form of birth control, for to restrict sex to the sterile interval is, in some sense, 'natural' (and as the technique has a high failure rate a steady influx of candidates for baptism is guaranteed). Judaism, too, is strict in its views about timing, perhaps in the interests of fecundity. During the actual menstrual period, 'she shall be put apart seven days: and whosoever toucheth her shall be unclean until the even'. It also enjoins that a woman should be kept separate from her husband for seven days after the flow has ceased (an interval in which she has almost no chance of conception). Then, after the ritual bath, she enters her most fertile period – and intercourse is recommended. Whether such rules come from observations of fertility patterns, or whether they just reflect a general obsession with cleanliness and the power of blood, nobody knows.

Women, unlike chimpanzees, advertise their potential for copulation at all times, fertile or otherwise. Perhaps a false statement of fecundity means that a male will choose to stick with a particular mate in order to keep others at bay, rather than to make a switch to a third party while his partner is unable to conceive. It may also help to ensure that the man will invest in the care of children which, he can be pretty sure, are his own. A more cynical view is that a female, more aware of her state than her husband, can have affairs with another partner but that any progeny will be brought up with the help of her regular mate. Yet another notion, unromantic but simple, is that once our ancestors began to walk upright bulges in the relevant region became an inconvenience and faded away.

Some claim that female voices, or faces, or dress sense, or body odour, or erotic judgement, change at the moment of maximum fertility. Women might be choosier – and more chosen – at those times, and on the crucial days could perhaps entice taller, richer, or more macho males. Men may, in retaliation, be able to detect quite where in the cycle a potential partner might be. A rigorous study of eleven lap-dancers found that women earned more in tips just after they had ovulated than in an infertile part of the cycle. All this might reflect an evolutionary race between males and females rather like that between parasites and their hosts, in which the female evolves to hide her reproductive state and the male comes up with improved mechanisms to see through her disguise.

Much of this is fascinating but fantastical and is infested with speculation, weak science and over-interpretation. The new genetics reveals more significant conflicts between males and females that rage below the surface, at the level of the DNA itself. They strike at the egg, the foetus, the newborn and even at later generations. Perhaps the most remarkable discovery of recent years has been to find that fathers and mothers fight not just over the fate of their own genes but over those of their progeny.

All cells contain a complete set of genetic instructions, but most remain unread in most places for most of the time. In some no more than a few sections of DNA are at work, while in others – like those in the brain – a great array is switched on and off as the machinery goes about its tasks. The human genome project did nothing more than make a list of parts. What really matters are the rules of assembly and the machinery of control, and they are intricate indeed.

A mother wants to invest just the right amount of biological capital into her child; too little and it may not thrive, too much and she may damage her own reproductive future. A father's needs are less subtle. He does not put much in to each sexual adventure and can always abandon his partner to find a new mate. As a result, male interest in the welfare of a female's later offspring (who might be the progeny of another man) is less than hers. It hence pays a father to persuade his infant to extract as much as it can from its own mother, however much that may damage her later pregnancies. Such asymmetry manipulates the economy of foetuses, of infants and even of adolescents.

The first hint of the molecular clash between men and women came from inborn disease. Just a couple of decades ago, it seemed that almost all inherited errors were passed on just as Mendel had observed with his yellow or green peas; one copy came from the father and another from the mother and it made no difference who transmitted what. The idea that events at conception could influence the next generation was dismissed as an ancient myth. Genesis, for example, speaks of Jacob, who tries to persuade his animals to produce new patterns in their offspring: he 'took him rods of green poplar ... and pilled white strakes in them ... And the flocks conceived before the rods, and brought forth cattle ringstraked, speckled, and spotted.' That notion was once mocked: but attitudes have changed.

Thirty years ago, the symptoms of certain rare and apparently distinct conditions were found to be due not to alternative forms of the same gene but to whether the same section of the double helix had come from the mother or the father. Somehow, the activity of the DNA was altered in the interests of the person through whom it was inherited. For the first such example, passage of the gene through the father generated an infant who was restless, stubborn and hungry while the same error inherited from a mother produced a child that was – conveniently for her – calm, passive, and satisfied.

Each parent puts a molecular stamp – an imprint – on the DNA that alters its activity in the fertilised egg and long afterwards. Often, paternal transmission increases a child's demands on the mother, while inheritance from the mother does the opposite.

The first three months of life are the only time in which growth depends only on how much food is available, for after then hormones control how tall or heavy a child becomes. In that crucial period, the molecular memory of one or the other parent plays a large part in an infant's fate. Babies in whom the paternal stamp of approval is prominent grow faster than do their opposites. Genes imprinted by a father do their greedy job in several ways. They increase the size of the muscles used to suck at the breast, make the tongue (a suction pump) larger, cause the infant to cry louder and longer and so increase the mother's production of certain hormones that help the newborn to gain her affections. The battle goes on into teenage years, for male-transmitted genes lead to a late puberty (which means that the child continues to demand attention from its mother) while those marked by a female do the opposite.

Experiments on mice show the real extent of parental selfishness. Before birth, maternal imprints tend to play the larger part, but after that the father makes his presence felt. Eggs with two female, or two male, genomes can be persuaded to develop up to a certain point. Those with two mothers put their effort into the embryo itself, while those with two fathers are much more interested in the placenta, whose main function is to extract food. Dozens of genes active in the mouse brain go through the process – and those marked by fathers are busiest in the sections associated with hunger. The battle between parents means that virgin birth in mammals (ourselves included) is impossible, for without the correct balance of genes from each of two parents the embryo will not develop.

For mice, men, and women alike, sex is filled with conflict, expense and anguish. How, in the face of such problems, does it maintain itself? Why indulge when the advantages of a chaste existence seem so obvious? The few exceptions to the rules hint at why the habit remains popular.

Some creatures have taken the asexual route. Plants copy themselves with runners, tubers or broken fragments that can generate all-female populations or go in for self-fertilisation, with the male and female parts of an individual hermaphrodite mating with each other. Animals as sophisticated as slugs, fish and lizards share some of those habits.

Efficient as asexuality may seem, it is rare. Of the two million or so kinds of animal known, just one in a thousand has taken it up (and some among them may have discreet liaisons not yet noticed by biologists, with males far smaller than females, or with an occasional fling just once in a few decades). Those who choose such a path are scattered through the grand tree of life as proof that each must have had sexual ancestors at some time in the past. They have fallen behind, for few asexual species have given rise to any descendant forms. In the long term, to abandon males has been a false economy.

Part of the reason has to do with mutation. Such events are infrequent, at around just one in a million per functional gene each generation, but there is so much DNA that every sperm or egg has thirty or so changes compared to those who made them. A few may be advantageous but many more do harm. As the mistakes build up, they lead in time to the decay of any lineage. All clones accumulate variation of this kind and as a result, sooner or later face nemesis. They also suffer from the matching problem that, as asexual generations succeed each other, the chances of any line picking up favourable changes, one after another, are very small. For such a lineage to do so it must draw a winning ticket in the biological lottery every time it reproduces.

Sex turns reproduction into a poker-game rather than a sweepstake. It makes a new mix of genes every time. As a result poor hands can be discarded, while any new and advantageous combinations that turn up can be put on the table. At the centre of sexual reproduction is the rearrangement of the inherited code of each parent into new combinations as segments of maternal and paternal DNA assort, break and rejoin. Before sperm meets egg there is a great biological reshuffling of cards; not — as in poker — with fifty-two items in each hand, but with millions.

When the DNA of two Englishmen, two Kenyans, or two Chinese is compared, about one letter in a thousand differs between them. They also diverge because of the presence or absence of vast numbers of insertions and deletions of particular segments, transfers among chromosomes, and reversals of the order in which the letters are arranged. The extent of diversity, constantly rearranged as it is by sex, is as a result almost beyond comprehension. Each generation, almost an infinity of new biological hands of cards is dealt into the game of life.

George Bernard Shaw was once propositioned by the actress Ellen Terry, who suggested that they should have an affair; their child, she argued, would have her beauty and his brains. Ah, said Shaw, but what if it has my beauty and your brains? As he had noticed, fornication is a two-edged sword. In the bearded, dim (and probably doomed) issue invoked by the playwright, two harmful attributes could be lost at the cost of a single death. In the other combination — Ellen Terry's child of bright visage and brilliant mind — advantageous features would be brought together for the first time.

When I was in my predictably gloomy teens I used to go, now and again, to the local art gallery, donated to his grateful soap workers by Lord Leverhulme, the builder of the model village of Port Sunlight. The place was usually deserted, as the founder's tastes, which tended to the Pre-Raphaelite, were in those days out of fashion. A certain work had particular appeal to adolescent angst. It depicted a thirsty goat with a bowed head, on the edge of collapse in a salty desert. It was painted in 1856 by Holman Hunt, who was much occupied with biblical themes and wished, so he wrote, 'to use my powers to make more tangible Jesus Christ's history and teaching'. The image is that of the animal described in Leviticus: 'And Aaron shall lay both his hands upon the head of the live goat, and confess over him all the iniquities of the children of Israel, and all their transgressions in all their sins, putting them upon the head of the goat.' On the frame is inscribed the remainder of that quotation: 'And the goat shall bear upon him all their iniquities unto a land not inhabited.'

The image is of the scapegoat, the sacrificial animal ejected into oblivion by the priests on the Day of Atonement. It is selected by lot, and dies in misery, laden with the accumulated sins of the Israelites, which are liquidated as a result. Holman Hunt saw in that a resonance of the death of Christ himself, which purged those who accepted Him of their multiple sins (his hope that the work would inspire Jews to convert was not fulfilled).

The message of *The Scapegoat* has a biological parallel, for sexual reshuffling can cleanse a population of accumulated weaknesses. Any new and feeble mixtures that emerge are heaped upon the shoulders of a metaphorical goat, a newborn infant. Should it die or fail to reproduce, its failure atones for the multiple faults present within the population. A single death purges many errors.

Many of our foes are themselves fond of sex and to fight

them off we need scapegoats of our own. Parasites and hosts are pitted in an endless contest in which neither party can afford to let down their defences. The scriptural link of sex with impurity is justified for it reflects a real tie with the diseases that have afflicted ourselves, our crops and our beasts since agriculture began. The Bible records the complaint by Joel: 'That which the palmerworm hath left hath the locust eaten; and that which the locust hath left hath the cankerworm eaten; and that which the cankerworm hath left hath the caterpiller eaten.' Such creatures have blighted the fields from Old Testament times to today.

The Irish potato famine of the 1840s was caused by a pest called the potato blight. At that time, the crop represented a third of the island's entire farm production. It was the staple of the Catholic peasantry, whose grain, milk, meat and eggs were exported to England or eaten by the Anglo-Irish in Ireland itself (which led a local to write that 'The Almighty sent the potato blight, but the English created the Famine'). A certain variety of potato, the 'lumper', was easy to grow and became widespread. In the 1840s came disaster. A frightful infection began to spread. Its agent finds its relatives among certain seaweeds. It travels at speed with airborne spores and is most destructive in cold wet years which slow the tubers' growth and render them susceptible to the autumn peak of the blight itself.

The 1840s saw a series of such seasons. Soon, three quarters of Ireland's crop turned into slime. Starvation roamed the land. Many people, unable to pay rent, had their houses pulled down around their ears and were thrown into the fields. A million died. The disaster was seized upon by anti-Catholic missionaries from England who offered food to anyone who would convert to their own sect, but most of the locals chose to suffer instead.

The epidemic came from a shortage of males. Potatoes in the wild often reproduce sexually, but on farms all are propagated through the asexual route, with tubers. In Ireland at that time, almost the whole crop consisted of a single clone, with every individual 'lumper' identical to all the others.

Unlike its targets on the farm, the organism of blight is sexual. The pest evolved in the same place as its victim. In the Andes, many wild varieties of potato, each filled with diversity, indulge in sexual reproduction as they flower and exchange pollen. In that dramatic landscape the agent of disease is in an evolutionary race with its hosts, who must reshuffle their genes at great speed to provide resistance against infection. As a result, the fungus can attack no more than a small proportion of its potential targets at any time. In a wet and cold Ireland, with millions of identical plants, just one of the vast diversity of blight strains that emerge each year thanks to sexual reproduction circumvented their defences and killed them all.

Potatoes still rank third on the global list of food crops and blight remains a real threat. It causes four billion pounds' worth of damage each year and farmers use vast quantities of chemicals to keep it at bay. In an attempt to reduce the risk of another famine, expeditions to the Andes and elsewhere in South America have searched for resistance genes in wild strains. They have found protective variants in native potatoes from Mexico, Bolivia and Chile; and a bank of their DNA is now maintained in a Dutch research station. Crosses introduce new protective variants into the crop from time to time in an attempt to keep ahead of the pest. The parasite's dangers have increased with its recent hybridisation with another form. That novel mix of genes enables the hybrids to attack distant relatives of the potato such as tomatoes and sweet peppers, and reminds us that sex is an ingenious, opportunistic, and risky business.

The approach used by potato-breeders is now applied to many crops – corn, maize, tomatoes and more. The process is expensive but shows that, in spite of the high cost of males, to get rid of them would lead to ruin.

The battle of parasite with host began long before farms were ever thought of and is as bitter in the animal world as among plants. A natural experiment in which I myself played a small part shows how fine the balance may be. In 1888 a ship from New Zealand anchored in the lower Thames to await a cargo. There it dumped gallons of freshwater ballast picked up in its home port. In it lurked some inconspicuous black freshwater snails called *Potamopyrgus*. The Antipodeans found the Thames to their taste and began to reproduce, without benefit of sex. They soon out-competed the various sexual species of native snail. Within a century they had spread in vast numbers across Europe, the United States and elsewhere. Thirty years ago I began to study them with the then primitive methods of molecular biology. British populations consisted, we found, of three distinct and all-female clones, each with its own set of genes and each differing in a small degree in where they lived.

The result had no discernible impact on biology (and a letter to Ken Livingstone, then Mayor of London, with a request that the local riverside park be renamed Potamopyrgus Park after the successful female immigrant was just as ineffective), but with my usual talent for missing the point I failed to ask the crucial question: what happens to the creature in its native land? Others soon uncovered a remarkable tale.

In New Zealand the snail is found in many places, from estuaries to mountain lakes. In the steamy lowlands the creature is infected by a variety of parasites, and around a quarter of the animals are males. In icy lakes, where such enemies are absent, all the snails are female. In the nineteenth-century Thames both sexes may have arrived, but in the absence of New Zealand's parasites the all-female forms prevailed. Now, rumour has it, some of their enemies have made it from the Antipodes to Europe and a new evolutionary experiment is under way, with a possible return of males to the virginal world of the migrants or, perhaps – as in Ireland – their utter destruction by disease.

For *Potamopyrgus*, potatoes and people, sex is an endless circular relay-race in which once the first player takes up the habit, all others are forced to follow or, sooner or later, to fail. Aaron, the elder brother of Moses, who utilised a scapegoat to 'confess over him all the iniquities of the children of Israel', presaged some very modern biological ideas.

That stern Old Testament view evolved into the central theme of the New: that those who accept Christ will be purged of their sins and find a new life in freshness and purity.

The notion is close to one theory as to how sex began in the first place. It emerged as a mechanism to renew the corrupted bodies of each parent and to produce rejuvenated versions that are, quite literally, reborn. St Peter prophesied that his followers, too, would have the luxury of 'Being born again, not of corruptible seed, but of incorruptible, by the word of God, which liveth and abideth for ever.' In the physical world that is over-optimistic for the infant will itself be tarnished by age, but to enter the world renewed even in temporary form is impressive enough. The double helix abides more or less for ever, even if those who transmit it do not.

DNA is a complicated molecule. To chemists, used to the challenges of building elegant but unstable compounds, such an enormous structure should be impossible to copy without it breaking down. Each time it replicates it makes many mistakes and if they were not put right the system would fall into chaos. Evolution has solved the chemical dilemma in its usual rough and ready way, and most of the faults are corrected with specialised proteins that edit out inaccuracies. Their efforts mean that the error rate for most of its letters is equivalent to copying out the Bible three hundred times before the first slip is made.

Like proof-readers, the enzymes turn to a reference copy to check what the correct message should be. Before the invention of printing, scribes, as they carried out their lengthy and wearisome work, needed an accurate version of the manuscript to check against their own production. Their modern equivalents have better aide-mémoire for they use a dictionary to check whether their spelling is accurate. DNA repair calls for much the same kind of help. Before the enzymes can make corrections, they must refer to a back-up to see what has gone wrong before they cut out the error and paste in the right version.

The mechanism of repair is rather like that of sex, for that process too cuts and rejoins the double helix. DNA repair might indeed be how sex itself began. In the earliest days creatures with single copies of their genes were forced to appeal to others to obtain essential editorial information as they patched up the damage caused by mutation. The first arrangements may have been close to cannibalism as one damaged organism swallowed another, less imperfect, individual to learn what was necessary to fix its own genes. That then edged into cooperation. A remarkable group of creatures hints at how this may have happened.

Bdelloid rotifers are tiny freshwater animals that row themselves through a liquid world. They can dry out into suspended animation for years, but will wake up again with added water. Bdelloids are widespread, abundant, and diverse, with more than three hundred different species spread across the globe. They are the great exception to the rule that asexual organisms have short evolutionary lives. No males have ever been found, and the fossils of animals trapped in amber suggest that the same has been true for forty million years.

One key to their survival is simple, for bdelloids have a clever means of escape from enemies. They dry up and blow away. When kept in water, sooner or later they are attacked by lethal fungi, but if an infected culture is desiccated for a few weeks and water is then added, they are miraculously restored to health, for the fungus cannot survive the drought. A period spent in the desert allows the animal to triumph.

Bdelloids also have a unique method of keeping ahead of decay. It has a message for biology as a whole. Over the generations the various clones have accumulated plenty of differences through mutation. Some of the millions of errors which have taken place in their long history were, no doubt, harmful. How could they be repaired without the back-up used by all other biological proof-readers? The answer is bizarre, for each generation the animals pick up random lengths of DNA from their dead fellows and from other creatures in the water. They patch up their double helix with the corpses of strangers.

That, perhaps, was the very first version of sexual reproduction, in which – in an echo of Holman Hunt's interpretation of the Crucifixion – the dead gave up their bodily selves to save the living. The process developed into cannibalism, in which members of one's own species were engulfed to obtain the necessary information. Later yet came the mitigated form of cooperation known as sex, in which two individuals of the same species get together to compare notes and put their mistakes right. That model is stark but speculative and the real truth about how life escaped from its clonal prison is still far from clear.

The moral implications of sexuality, on the other hand, are still, as in biblical times, a matter of confident assertion. In 2001 Pope John Paul II criticised George W. Bush with these words: 'A tragic coarsening of consciences accompanies the assault on innocent human life in the womb, leading to accommodation and acquiescence in the face of . . . proposals for the creation for research purposes of human embryos, destined to be destroyed in the process.' A month later Bush banned all federal support of research on clones of stem cells derived from embryos, apart from upon a few lines already cultured in the laboratory.

In 2011, to general astonishment, the European Union's highest court, the Court of Justice, agreed with him. They ruled that processes and products based on human embryonic stem cells cannot be patented because their use 'offends human dignity'. Their definition of those whose *amour-propre* might be affronted includes any fertilised egg, but also an unfertilised egg that has been persuaded to divide, and those with two male, or two female, cell nuclei (neither of which could ever develop into an infant). In the hallowed halls of Strasbourg, with plentiful advice from the Church, the ancient argument about clones has been brought back to life and the soul now has the full protection of European law. What my mother and her twin might have thought about that, I will never know.

CHAPTER FOUR

ON TO METHUSELAH

William Blake, *Marks of Weakness, Marks of Woe*

But if a man live many years, and rejoice in them all; yet let him remember the days of darkness; for they shall be many.

ECCLESIASTES 11:8

Methuselah did best, at nine hundred and sixty-nine, but his predecessors met their end almost as full of years – Adam at 930, Adam's son Seth at 912, his grandson Enos at 905, and so on until the rot set in with Methuselah's own son Lamech, who passed away as a stripling of just 777 (Methuselah's father Enoch had departed this Earth at no more than 365 but not to die, for he 'walked with God' instead and begat sons and daughters as he did so). Alarmed by man's desire to linger so long in the temporal realm the Lord then decreed that 'His days shall be an hundred and twenty years.' There has been a further falling away from his commands since then.

Lives in those patriarchal days were long but they were also consistent, with a difference in survival of less than one year in ten between the oldest and the youngest among the elders. So predictable was the appearance of the Grim Reaper that the age at which each breathed his last might have been set into his body at birth.

Aristotle, too, claimed that there was an inborn date with destiny. He likened our time on this planet to a blazing fire. A baby was born with a dose of vital force. Death came in two flavours: premature, with the flames extinguished when water was poured on them, and programmed, when the blaze burned itself out. The search for the fuel of life has continued since his day. Aristotle's metaphor works well for, like a bonfire, each of us is a chemical reaction that is bound sooner or later to run out of raw material. What determines when it will gutter into extinction?

For most of history few such fires had the chance to flicker out of their own accord. Men and women, boys and girls, people in their prime and those advanced in years faced a metaphorical dash of icy water at each moment. Many died as infants, some as young adults and no more than a few reached ripe maturity. The accidents of existence meant that survival at any age was always far from certain. Until not long ago, the loss of a child was more commonplace than is that of a parent or a grandparent today. Rich or poor, town or country, male or female; for all of them violence, infection, accident and hunger meant that the Angel of Death might arrive at any moment. His presence loomed, as a result, rather larger in past imaginations than it does in our own.

In biblical times average life expectancy was not much more than around thirty (and much of that came from infant mortality) and every moment was filled with threat. In the modern world, in contrast, most people can be confident about when they will breathe their last, not because – as in the days of the Patriarchs – the old die older but because the young die old. We are in a new Age of Methuselah, not in years but in consistency. Today's funerals are as easy to foretell as were those of Genesis. Almost all infants born in richer countries can now be confident that they will walk the Earth for three score years and considerably more than ten. Since Old Testament times the shape of death has been transformed.

Every nation wants to count its people. In the Book of Numbers the Lord commands Moses to 'Take ye the sum of all the congregation of the children of Israel, after their families, by the house of their fathers, with the number of their names, every male by their polls.' That first census took place 3450 years ago. Moses found 603,550 adult men able to go to war (the Tribe of Levi would have boosted that figure but its members were excused combat on the grounds that their job was to service the Tabernacle). His successors across the globe have, to raise armies or pull in cash, totted up their subjects again and again. The tax-collectors who wrote the Domesday Book were thorough: 'There was no single hide nor a yard of land, nor indeed one ox nor one cow nor one pig which was not set down.' They found that the population of England (or that of the parts surveyed) in 1086 was rather less than two million.

The first census-takers had no interest in anything but the basic figures. That changed when their data were used by the haberdasher John Graunt to tease out London's patterns of birth and death. In his 1662 book *Natural and Political Observations Made upon the Bills of Mortality* Graunt devoured the columns of numbers to 'present the World with real fruit from those ayrie blossoms'. The 'blossoms' were the records of burial and baptism (the 'bills of mortality') maintained by parish clerks. The 229,250 burials recorded over twenty years did not state the age at which each person perished, but for thirty years deaths had been recorded with a cause of demise, as identified by 'searchers', ancient matrons paid to make post-mortem diagnoses. The afflictions included ague (malaria), apoplexy (stroke), chrisom (death in the month after baptism; so called after the baptismal robe upon which the corpse was rested), frighted (heart attack), overlaid (an infant smothered), strangury (kidney disease), rising of the lights (croup), together with an unfortunate 'bit with a mad dog' and another who perished of piles. Apart from the dog and perhaps the piles, each agent tended to attack those of a particular age. They could hence be used to work out the chances of survival at different stages of life.

Many of those who appeared in the Bills had 'died of the

thrush, convulsion, rickets, teeth, and worms, or as abortives, chrisoms, infants, liver-grown, and overlaid that is to say, that about ½ of the whole died of those diseases which we guess did all light upon children under four or five years old'. Chrisoms and infants were the largest category, well ahead of those who expired of consumption (now called tuberculosis). One child in three met its end before the age of six and just one in ten of the survivors to that age made it to the biblical three score and ten. For those who survived the horrors of infancy the rate of mortality increased with each year that passed. The picture was clear. London was a dangerous place, for children most of all, and for those lucky enough to become adult each successive year was riskier than the last. John Graunt had made the world's first 'life table'. It enumerated the chances of demise at different ages.

Mortality underwent great swings in the years of his survey and in the two decades of his main analysis just eight perished of the Black Death, while in the Plague Year of 1665, when he was still busy with the figures, 68,958 of the 97,036 burials were from that disease. Graunt presented his book to the new Royal Society and in spite of some reluctance to welcome a haberdasher to its ranks he was, after pressure from the King, elected a Fellow.

The pattern of mortality in Enlightenment London marked many places for many years. Most of the time, medicine could not help. Often, it made matters worse; as an eighteenth-century poem had it: 'One physician like the sculler, plies,/The patient lingers and by inches dies./But two physicians like a pair of oars,/Waft him right swiftly to the

Stygian shores.' In France a hundred years after *Bills of Mortality*, Rousseau claimed that 'Half of the children will die by eight years of age. This is an immutable figure. Do not try to change it.' His advice was not taken.

The first full British census was held in 1801 and with the exception of 1941 the count has been taken once a decade since then, although the 2011 round may in this electronic and parsimonious age be the last. The figures paint a picture of huge shifts in survival. As the capital grew to become the biggest city in the world its children had an even worse time than in the statistical haberdasher's day. The rate of loss reached a high point a century after Graunt (perhaps because of the arrival of new and virulent strains of diseases such as smallpox as trade increased). Then it fell fast. Two out of three London children expired before they were five in the 1750s and no more than half that proportion a hundred years later.

Progress continued, thanks more to the efforts of architects and engineers rather than of doctors. Almost half the drop in mortality in the nineteenth century, which showed a gain of more than a quarter-century in life expectancy, was due to a reduction in tuberculosis as people moved from the insanitary and crowded slums. Most of the remainder came from the defeat of louse-borne typhus, of typhoid and cholera carried in dirty water, and of dysentery due to unclean food. In those days, even the children of doctors had death rates not much different from those of the poor, for almost nothing was known of the causes or treatment of disease.

Now the picture has changed again, for the youngest most

of all. In the United Kingdom, infant mortality, the loss of those less than twelve months old, has dropped to one in two hundred (many among them victims of inborn disease), a fortieth of what it was at the outbreak of the First World War, when diarrhoea caused by contaminated milk or water was still a scourge. Antibiotics, vaccines and other medical advances began their inexorable progress after the Second World War. For those under five, Britain's death rate is down to between six and seven in every thousand live births, which is impressive but remains above the Western European average. The twenty-first century is a good time to be a baby.

The old too have reason to celebrate. The Genesis promise to Abraham that '. . . thou shalt go to thy fathers in peace; thou shalt be buried in a good old age' has for most people been met. The Third Age has mounted a take-over bid. For the first time, the number of Britons over forty-five is greater than those below that age. In fifteen years, people like me – obliged to retire while still in good order – will be the most frequent class in the population, an honour that now belongs to those in their forties. Most will look (and perhaps feel) more youthful than I do today.

Existence in the modern world, frenzied as it might seem, is in truth slower than it has ever been. People stay in education for longer, are supported by their parents into their twenties, remain healthy for many more years, and start their families much later than before, with the average British woman delaying her first child until the age of thirty, five years older than in the United States (and matched in sexual reluctance only by the Germans). As a result the inhabitants of these islands have fewer children than in earlier times with an average of slightly less than two per couple in 2012, while it fluctuated at around three in the 1950s; a shift which accounts for some of today's abundance of the elderly. For the first time in history, most parents live to see their grandchildren and most women pass half their lives in an infertile state. Even so, the inexorable increase in the chance of extinction in each extra twelve months of adulthood uncovered by John Graunt has remained.

What is age? Some animals manage without it. Their risk of death does not increase with time. They are like pint glasses in a bar. Each has a chance of breakage on any given day, but the odds do not depend on how old they are. A brand-new tankard is at just the same risk of disaster as is one bought twenty years earlier. The patterns of mortality of the tiny freshwater animal *Hydra*, a relative of jellyfish, are of this kind (even if the vast majority die young, of simple bad luck). For most creatures, in contrast, the likelihood of obliteration goes up with time. For men and women, after infancy the risk of death in a particular period doubles every eight years. As a result, people who celebrate their hundredth birthday can get no more than an even bet that they will blow out the candles on their next. The figures are more favourable in the prime of life, from twenty to forty or so, and are at their best at the age of ten. If that schoolboy rate of demise were to persist most children born in 2000 would survive until the year 3300, which knocks Methuselah into a cocked hat and gives an uncomfortable insight into the power of decay.

Today's rhythms of mortality mean that more and more

among us face the prospect of prolonged decline. In spite of the well-attested survival of the Frenchwoman Jeanne Calment to 122, most claims of extreme age do not hold up. Some countries have more success than others, but part of their lead is due to relatives who continue to draw pensions for people who passed away long ago. The Russians once issued a postage stamp to celebrate the 148th birthday of a citizen, while Colombia beat that with an issue devoted to a 167-year-old. Both issues represented self-congratulation; national philately out of control.

Even so, a few places do stand out in the survival stakes. The people of the Japanese island of Okinawa last to eighty-four, four years longer than most of Europe. Those on the Aegean island of Ikaria are almost three times more likely to make it to ninety than are Americans while groups such as Seventh Day Adventists in the United States also have unusually lengthy lives.

Some boasts of extreme antiquity may be unfounded, but the real figures are remarkable in their own right. The developed world is in an era of death postponed. The average life expectancy for British men in 2012 was 78.2 years and for women 82.3. The average was, once again, well up compared to the previous year. Since I emerged into this world on the day of the Great Escape from the Stalag Luft III prisoner of war camp (March 24th 1944, for those who need to be reminded) the average age at which a citizen can expect to survive for another ten years has risen by a decade. In 1944 it was seventy, but is now close to eighty. At the end of the war, an eighty-year-old could expect to see just five more summers, but now the same days of sunshine await people of ninety. These islands have four times as many centenarians than they had in the 1960s. In 2012 the total was over twelve thousand and at the present rate of progress it will hit a hundred thousand before the end of the present century.

Even better, the era of 'sans teeth, sans eyes, sans taste, sans everything' – of debility in the last years – has been delayed. Today's sixty-year-olds have the health of people twenty years younger at the time of the First World War, and our octogenarians that of people of seventy when the Beatles made their first LP. By 2300, says the United Nations, one person in three, worldwide, will be over sixty-five and a typical citizen of a developed nation will last until he or she is a hundred. Nowadays, assuming that sixty-five is the universal age of retirement, a typical pensioner, averaged over the globe as a whole, has about two weeks in which to enjoy his or her gold (or plastic) watch. By 2300 he or she will have, if the UN is justified in its optimism, thirty years.

Since the middle of the nineteenth century longevity in the developed world has gone up at the extraordinary rate of six hours a day, unchecked through peace and war, boom and slump. Across Europe, nations have come together in the patterns of survival. As recently as 1960, Portuguese men died seven years younger than did their fellows in Spain, but now the difference is two years (and Spain does better than Britain). If such progress continues, most children born in Western Europe since the millennium will see in the twenty-second century and a few might make it to a hundred and twenty.

Some people suggest that even that might be an underestimate. Demographers have often failed to forecast the rate at which lives will lengthen. As recently as 1977 they estimated that a British male born in 2010 could expect to survive to 71, but that figure has already risen by almost a further decade. Optimists claim that the first person to celebrate his or her thousandth birthday has already been born.

Pessimists insist, with good reason, that this is complete fantasy. Even so, many doctors suggest that today's treatments for the major killers strike so hard at the roots of age that there is a real prospect of delaying its advance by another ten years.

Can we continue to postpone the fate that awaits us all or are we – like the Patriarchs – set to expire after a programmed stay on Earth, however well we live? In this less than perfect world, how much closer will men and women get to Methuselah, or even to God's permitted span of a hundred and twenty? How much of today's improvement is due to social and medical progress, and how much, if any, to a real shift in the rate of decay?

Life is a biological race against time. As always, what biology can do is limited by physics and chemistry. To claim that recent advances make it likely that most people will now reach their century is like saying that, given the improvement in the mile record from four minutes in 1954 to its present three minutes forty three, in the near future athletes will run the distance in only three minutes (or, for that matter, that in two millennia they will travel at the speed of light). There has in fact been little change in the record for a decade as a hint that a biological limit is not far away. Is the same true for the race from cradle to grave?

Whenever nature's fire-extinguisher puts out the Aristotelian blaze, most people in the developed world a century from now will be doomed, as they are today, to die like Abraham, 'well stricken in age', of decay. Smallpox has gone and leprosy may be next, while the drop in the numbers of smokers means that lung cancer is less common than it was. Accidents still kill, but seat belts have helped and few perish from cold or hunger any more. Medicine and clean water have played their part but wealth, at a level unimaginable even half a century ago, brings warmth, food, relaxation and education (itself a powerful predictor of longevity). Threats from outside still put paid to some before their time, but for most of those who read this book the main menace remains their inborn inability to repair their own machinery.

Not everyone has reached that happy state. Britain may lag behind Western Europe, but Indians die ten years younger than we do. The USA was for much of the twentieth century a leader, but its modern inhabitants last a year less than even those of the United Kingdom. Without the reduction in mortality since 1900, the nation would hold no more than half as many people as it does today. Even so, in the past thirty years it has lost seven million who would have survived had its society kept up with those of its rivals. For some of its citizens things are getting even worse. The life expectancy of the least-educated white Americans has dropped by four years in the past two decades, and the poorest white women now live a year less than their black equivalents. The USA sank from the middle of the survival table for females in developed countries in 1950 to last place in 2010. Other places have done even worse, for they have not progressed much beyond John Graunt. Infectious diseases – AIDS, malaria, tuberculosis and more – kill off multitudes. The bottom ten nations are all in Sub-Saharan Africa, with Chad in such a desperate state that the average citizen sees no more than forty-seven summers.

There, as in Enlightenment London, the young suffer the most. In Angola infant mortality reaches the deplorable rate of one in six. A third of humankind is in their teens or twenties. Nine out of ten of those young people live in the developing world, and they include almost all who die before their time. Disease and hunger kill some but violence accounts for more. Young men destroy themselves at twice the rate of their sisters, in car accidents and with murder and suicide close behind. Women expire in thousands because of lack of care in childbirth. Botched abortions are another killer, with mortality a hundred times higher in some countries than in others (and in South Africa, the numbers fell by 90 per cent after legalisation). The World Health Organisation estimates that seven out of ten such deaths could be avoided. Gun control, seat belts, legal abortion and even the control of certain pesticides (used in India to lethal ends) could all help.

Even within these islands, the contrasts in survival are stark. Glasgow has the lowest male life expectancy of any city in Western Europe, and it is getting worse. It possesses a magnificent Victorian cemetery, the Necropolis, which was founded in 1832 in an attempt to emulate the famous Père

Lachaise in Paris and, *post mortem*, 'to afford a much wanted accommodation to the higher classes'. In that it succeeded. The place now boasts three and a half thousand memorials while ten times as many of its denizens have no stone to mark their passing. The height of each monument (and some tower over the visitor) tells the tale of wealth and death a century and more ago, and reminds us how little has changed. The bigger and costlier the memorial, the longer, the inscription tells us, the occupant lived. Many among the proletariat died young and are buried with no hint of where they lie, while a good portion of the rich lived out their three score years and ten.

A trip to the Necropolis today is a chastening experience, for many of its bronze plaques and images have been stolen by the poor, now killed by drugs and alcohol rather than by infectious disease. Within modern Glasgow, men in the Calton district (a place of great poverty) die a remarkable twenty-eight years younger than those in affluent Lenzie, just five miles away (and Calton's male life expectancy is a decade lower than that of India). When I was a student in Scotland in the 1960s the nation's men lived, on average, just a couple of months less than did their English neighbours; today, the difference is two years. There has been no improvement in those figures for a decade and on present trends the gap is set to become wider.

Even for the people of suburban Lenzie, not everything in the memorial garden is rosy. Wealth brings its own threats. 2012 marked the five hundredth anniversary of Ponce de León's search for the fountain of youth which was, he thought, in what is now Florida. That land has a lesson for those who wish to delay the visit of the Grim Reaper. 70 per cent of its men and half its women are obese. The British, too, tip the scales further than do most other Europeans, with the Scots fatter than most (but not as fat as the Welsh). For smokers, drinkers, the idle, the stressed and the depressed much of the future lies in their own hands.

Be that as it may, many Scots and others live lives healthy enough to overcome most obvious external agents of doom. They must still face the enemy within, the biological timer inside every living being. To track down the physical causes of age might, as a result, slow the malign effects of the years.

What is remarkable is not that we age, but that we stay young for as long as we do. Each day every one of us replaces billions of cells and makes thousands of miles of DNA. The evidence comes from the Cold War. For twenty years after 1945 radioactive carbon was pumped into the atmosphere by atomic tests. Over time it has broken down but the element is still incorporated into our bodies. To track its turnover shows how fast each tissue is replaced. Skin cells last two weeks, the red cells of the blood four months, and a liver never reaches its first birthday. All of us are, taken organ by organ, younger than we think.

In that rebirth lies our fate, for the process of replacement is far from perfect. The decayed state of the aged, from grey hairs to cancer, reveals how inefficient it must be. As the years go by, many parts of the body begin to fail and, in time, the faults build up until a threshold is reached and the hearse appears at the door.

Our fellow creatures vary widely in when that moment

arrives. Invertebrates such as clams (whose age can be dated from growth rings in the shell) may survive for four centuries while plenty of insects last just a few days. Even mammals can rack up impressive numbers of summers. A bowhead whale caught in Alaska in 2001 was found to have a harpoon tip of a type not used since the nineteenth century embedded within it. Wild mice, in contrast, expire in just a few months. That two-hundred-fold difference in survival has evolved since mice and whales last shared an ancestor a hundred million years ago.

Length of life has somewhat of a fit with how risky the environment might be. Mice, whose days are uncertain, die younger than do elephants or whales, who are (or were) masters of much of what they surveyed. Individual differences within a species are also involved. Fruit flies and mice can be bred over many generations to have long or short lives, as proof of the existence of genetic variation in the rate of decay (although there remains plenty of random noise, for inbred populations in which every individual is genetically identical still have plenty of variation in life expectancy).

Men and women, too, have their fate programmed at least in part into the double helix. Identical twins pass away at ages more similar to each other than average, and the figures suggest that about a quarter of the variation in British lifespan reflects genetic variation. For small babies, DNA can be crucial, for many of those who die do so because they have an inborn abnormality. The importance of genes then drops over the years, but rises again in old age, most of all among the few who become very elderly indeed. A study in the Dutch town of Leiden examined that tiny set of families in which at least one brother or sister of someone of ninety or more had also reached that age. All those individuals, together with their deceased parents and sibs, and their children, had a mortality rate a third below the average, with fewer cases of diabetes, heart disease and cancer than did their husbands and wives (who had shared more or less the same environment). Genes become even more important in that tiny group who make it to a hundred and ten. For them, rather than an increase in their chances of departure each year as is true for most of us, their likelihood of survival remains at about 50 per cent for every extra twelve months that pass.

For age, as for everything else, nature and nurture work together. A certain genetic tendency towards premature death hints at just how complex their joint effects on decay may be.

Sex and death are close relatives. The gene that makes men male has the same effect on their exposure to the Grim Reaper as it does on their ability to run marathons and commit crimes. Husbands, with their toxic dose of testosterone, die younger than wives. Four women make it to their centenary for every man that does so. Testosterone damages the tissues and suppresses the immune system, so that those afflicted with it are less able to withstand infectious disease. Suicide, murder and poverty, too, are more of a risk for males. Even relaxation can be dangerous. Men are struck by thunderbolts at three times the rate of women, in part because they sometimes stand on golf courses with lightning conductors in their hands.

The malign effects of the Y chromosome on survival

depend to a large degree on the environment, be it ghetto or golf-course, in which it finds itself. Nowhere is this clearer than in the contrasts in patterns of male and female mortality between Western and Eastern Europe. Life expectancy has risen steadily in the West for both sexes, while women in the East have at least stayed more or less where they were; but the men of the former communist nations now do much worse. Up to the age of fifteen or so, boys and girls in Eastern Europe last on average only a month or so less than those in the West. Then, for men, things change. Over the past half-century, there has been an improvement in health and survival in middle age (from thirty-five to sixty years old) for both sexes across Western Europe and, marginally, for women in the other half of the continent. In the East, husbands have not kept up. There has been no overall improvement in their survival at all since the 1950s and some nations have seen a collapse in their health – Ukrainian middle-aged men have, for example, on average lost two years in expectancy since the 1980s. Those blessed (or cursed) with a Y chromosome have coped far worse with the new capitalism's problems of drink, depression and social disorder than have their partners.

That section of the genome has a large effect, but plenty of other genes also influence the rate of decline. Within many species, size matters. Large breeds of dog die younger than do small, and giants such as Irish wolfhounds expire at around six, at half the age of a typical miniature poodle. In the same way, a dwarf mutation in fruit-flies increases the animals' ability to survive. Many members of a certain family from a remote Ecuadorian village are less than three and a half feet tall as the result of a mutation in a gene in the hormone pathway that controls growth. They last for much longer than do their neighbours and stay almost free from cancer and diabetes. Even within the normal range, size makes a difference. Recruits for the Union Army at the time of the American Civil War in the 1860s were weighed and measured and followed by the authorities for the rest of their days. After taking into account their social status, and whether they lived in cities or the country, every extra inch in stature led to a 2 per cent annual increase in the chances of mortality.

Across species, the opposite pattern often holds. For most land mammals (bats excepted) each doubling of body weight leads to about a 15 per cent increase in length of life, which explains the mouse-elephant contrast. We ourselves stand outside that rule, for in spite of our modest stature we persist for longer than any of them (elephants included). Another unexpected zoological Methuselah lives underground in Africa. The naked mole rat's name describes its delightful appearance. It lives in subterranean colonies of up to three hundred. A single female does all the breeding for she bullies the others and suppresses their sexual cycles. The mole-rat is about the same size as a mouse, but lasts eight times as long – and stays young, healthy and fertile for two thirds of its days (which is the equivalent of an eighty-year-old woman having the biological make-up of a typical thirty-year-old).

What lies behind such vast differences in patterns of mortality? For ourselves, some causes, like smoking and alcohol, are obvious. Long-lived peoples such as those on Okinawa or Ikaria tend to have a restricted diet, with little meat and many vegetables and fish (the Greeks also use local herbs which contain substances that may reduce high blood pressure). Seventh-Day Adventists, too, with their extra decade on Earth, take heed of the instruction in Genesis: 'Behold, I have given you every herb yielding seed ... and every tree, in which is the fruit of a tree yielding seed; to you it shall be for meat' and insist on a vegetarian diet, ameliorated with small amounts of cheese and eggs.

Effective as such habits may be in delaying the return to dust, biology has not had much success in tracking down how they work. Even so, a few candidates for long-term survival have emerged and some have been used in attempts to delay the date of death, in laboratory animals at least. There have been many promises but the search for the Ponce de León pill – the panacea for age – has not yet succeeded. Its failure has a long history. The Chinese claim that foodstuffs that look like a man, or are red, will do the job. Gold, which does not degrade, also guarantees a longer stay on Earth while peach-flavoured Longevity Tea is popular in the United States, where Roy Walford's *The 120-Year Diet* was once a best-seller. In another approach, King David in his last days called for the great beauty Abishag the Shunammite, who 'ministered unto the king' (the process involved a transfer of heat) to prolong his days. Science is trying, with mixed success, to do the same.

At present, the best on offer is mitigated optimism. In the United States, where the average span rose by more than half during the twentieth century, the figure has gone up by only a fiftieth in the twenty-first. To improve it to eighty-five – just two years more than in today's Japan – would demand the elimination of mortality through cancer and heart disease. Other nations that do not suffer from America's poor health-care (which could in principle easily be improved) may find the job even harder. Even so, plenty of scientists have been happy to try.

The harmful effects of starvation or glut – both of which alter patterns of survival – are much influenced by inherited variation. Some genes respond to how much food is around while others protect cells against attack by the toxic by-products made when the stuff is burned. Yet others act as control valves in the body's economy and stop, slow or divert the traffic of energy. A related group deals with stress. When times are hard they put the economy on a war footing and switch on protective proteins. Defence becomes a priority, and sex must wait, which can lead to a longer (if not a happier) existence. To disentangle such mechanisms might be the key to our decline and, perhaps, to its defeat.

Many of the symptoms of the elderly, from wrinkles to heart disease, were once thought to emerge from the malign effects of a modified form of oxygen generated when food is broken down. The stuff wreaks havoc within cells and can cause inflammation (and many illnesses, from arthritis to heart disease, are associated with that problem). It also persuades cells to commit suicide before their time. However, the notion that we are poisoned by our own by-products is less popular than once it was. The naked mole-rat, for example, makes lots of activated oxygen but that does not interfere with its extraordinary span. Some claim that all warm-blooded animals (except for ourselves, who do much better) live for about the same number of heartbeats – around a thousand million – because that figure reflects the rate at which their machinery runs and hence the amount of waste produced. The hearts of many small and active creatures do flutter and their bearers do die young, but size is not everything. Birds beat mammals into the ground. A pigeon burns its daily bread at the same rate as does a rat and its heart beats at the same speed – and in flight much faster – but it lives ten times as long.

The reactive oxygen idea suggested to some that it might be possible to improve an animal's prospects by providing less fuel. It was found in the 1930s that to starve some creatures makes them live longer. In the laboratory at least, that simple trick works on worms, flies, rats and mice. A substantial reward, the Methuselah Mouse Prize, awaits those who most extend their subjects' days and a hungry laboratory mouse has already made it to more than twice its fellows' two-year average. For mice, even to start a fast in middle age improves survival – and in some creatures the mere taste or scent of food shortens lives (and the aroma of a good meal increases levels of the hormone insulin in humans, too). Starvation does not always work. Some mouse lines do not respond, and two twenty-year studies on rhesus monkeys gave contradictory results. In one, a cut of a third in food intake made no difference to longevity. In the other, there did seem to be a positive effect but the food used there was so rich in sugar that almost half of even the control group developed diabetes and all the experimental animals were themselves fat before the diet began. Laboratory rodents, too, are a pampered (and often overweight) breed. Hunger may do no more than help mice or monkeys shed a few ounces with the same happy effects as those experienced by humans on a diet.

For the obese, to cut down on food can indeed help them live longer. Whether it helps those of average size is less clear. Like the people of Okinawa and Ikaria, Westerners who reduce calorie intake have more good cholesterol, lower blood pressure, fewer reactive oxygen molecules, and less inflammation. Even better, their bodies work at a slower rate than before. All these changes, in animals at least, increase longevity, but for men and women the evidence is not yet in. It seems to pay to be of mean weight, or slightly below, for those much skinnier than average do no better than a typical citizen and may do worse. Athletes with their toned bodies do last three years longer than most of us. To concentrate on quality rather than quantity – to stay away from toxic junk – may make more of a difference. Charred meat and soft drinks fail on that score, and the food of chimpanzees or wild mice is also filled with harmful chemicals, many made by plants anxious not to end up as a meal for a hungry mammal.

There lies the secret of the mole-rat. In its subterranean existence, all food and water comes from roots, tubers and bulbs. They are packed with protective substances to fight off the attackers. Even daffodils, irises and crocuses have bulbs that cause vomiting and worse, while garlic and onion depend for their flavour on a reduced dose of the same stuff. The bulbs of African deserts are even more toxic but the mole-rat has a series of biochemical defences that deal with them. Its cells have become so tough that in culture they can cope with doses of lead, of cadmium, of drugs, of heat and of hunger that would kill off those of mice. Cells from long-lived birds and bats and other mammals (men and women included) are also resistant to such stresses, so that the ability to cope with poisons may in part be behind longevity itself. The mole-rat, for example, is almost free from cancer, perhaps because it can break down carcinogens. Whether the creature has lessons for ourselves, or whether it is no more than a zoological curiosity, we do not know.

Other genes change their activity when food is short. Some code for parts of the pathway that includes insulin, the foreman of the energy factory. As evidence of their importance they have almost the same structure in creatures as distinct as worms, flies and humans. A few come in alternative forms within a particular species, some of which increase in frequency as animals are bred to live longer.

Insulin itself persuades the liver to pump out a second hormone, an 'insulin-like growth factor' that controls the extent to which children grow. If the substance is injected into Ecuadorian infants with the dwarf mutation they become taller than their siblings. The factor, or another much like it, is found in mice, in fruit-flies and in worms, and when it has been damaged by mutation such creatures live longer. Small (and long-lived) dogs such as terriers have less of the hormone than do their gigantic relatives. Some very old people, too, have reduced levels.

Many attempts have been made to use such insights to develop drugs. Some claim that foods rich in substances that soak up poisons guarantee a long life, but people who swallow a variety of vitamins, or selenium, or carotene (all of which do that job) find no benefit. Indeed, those who take them are at a higher risk than average of premature death. Even a diet rich in fruit and vegetables does little to put off the evil day.

The 'French paradox' – a nation full of years in spite of its fatty food – was once ascribed to St Paul's advice to the biblical Timothy to 'Drink no longer water, but use a little wine for thy stomach's sake and thine often infirmities.' The stomach is said to respond to the joys of red, but not white, wine. A certain constituent of grape skins (left in the mix for the former but not the latter) has been claimed to improve survival in yeast, worms, flies, fish, and mice but that now looks less than clear. The substance has not been shown to work in humans and for a mouse to profit it would have each day to force down as much as is in a dozen bottles of Bordeaux.

Pills that persuade the body that it needs less food reduce the rate at which glucose is broken down and, in rats at least, cut down heart rate and insulin levels even as they increase the concentration of stress proteins. In worms they improve longevity but in small mammals do more harm than good because they damage the heart. Excess sugar does some of its harm when it combines with proteins to make a sticky brown substance that blocks arteries. A drug that protects against it gives mice a few more weeks of existence but, alas, causes stomach problems in humans.

Rapamycin is used to suppress the immune system after a transplant. Its name comes from the native term for Easter

Island, Rapa Nui, and the chemical is the product of bacteria found in the island's soil. In the laboratory stuff slows cell growth. Not long ago, to universal surprise, it was shown to increase the lifespan of mice by as much as an eighth (which, in human context, would mean around a decade). In older animals it did in proportion even better, with an increase of around a third. The substance interacts with a particular protein, part of the insulin-related chain of signals. When targeted by rapamycin, cell growth slows and less fat builds up, tumours reduce their growth and bone density drops at a lower rate. The drug also cuts down the rate at which misshapen proteins accumulate in nerve cells and cause conditions such as Alzheimer's disease. It may work on a crucial step in the mechanism of age.

Rapamycin is remarkable stuff, but it might dispose those who take it to diabetes and to increased blood cholesterol. A related drug, metformin, has been used without problems by millions of people with diabetes to reduce the amount of blood glucose. It interferes with the same pathway and might be a better candidate as an anti-ageing compound. Tests are under way.

The drugs may work one day, but most routes to a healthy old age are already obvious, for they involve little more than common sense; aged men, in the advice given to Titus, tend to be 'sober, grave, temperate, sound in faith, in charity, in patience'. All those attributes are, I am told, attainable by those able to make the effort. When it comes to life eternal, or even prolonged, such hints are helpful, but the new insight into the timers that control the fate of every cell hints that to get back to Methuselah rather than just escaping from the dangers of our own imprudent behaviour may be harder than it seems.

The problem emerges from the unholy trinity of sex, age and death. The body's ability to restore itself is – almost – infinite. The enzymes that scan the DNA correct the vast majority of mistakes made when the molecule is copied, but some do get through. Age is a reflection of the decline of the repair machinery, birth an affirmation of its renewal.

Once again, biology finds an echo in the Bible. The soul plays a large part in the Good Book's narrative, but undergoes noticeable development as the tale goes on. In the Old Testament it often refers to little more than individuals ('Whatsoever soul it be that eateth any manner of blood, even that soul shall be cut off from his people') but later it becomes the agent through which our flesh is reborn in the world to come: in the words of St Matthew '... fear not them which kill the body, but are not able to kill the soul: but rather fear him which is able to destroy both soul and body in hell.' Sex is much the same. It means, in the words of St Paul, that '... this corruptible must put on incorruption, and this mortal must put on immortality'. His intellectual descendant, Gregor Mendel, had much the same idea. For the Disciple the renewal of the soul on Earth allows our ephemeral flesh to be reborn while the botanical abbot proved that the genetic message is disengaged from the brief lives of those who bear it. In the world of body rather than spirit, sex rather than salvation resurrects the enfeebled flesh of one generation in the form of a new and healthy infant. Both processes involve death followed by rebirth, one literal or the other no more than metaphorical.

Sometimes the link is clear. Some birds and bats survive for years but they have few progeny, while starved fruit-flies live longer but cannot reproduce, while those forced to lay as many eggs as possible die young. In some creatures the tie is stark, for the male marsupial mouse copulates just once and then drops dead.

Many people have suspected a tie between sex and decline. Among the Huli people of Papua New Guinea boys are warned of the dangers of intercourse, which should not be indulged in except when children are needed. Some Indian sages insist that to produce a tablespoon of semen requires forty kilograms of food while in Victorian Britain loss of that vital fluid was said to knock off years of life. In the first decades of the twentieth century the Steinach operation – a slice across the tube that conveys semen to the outside – was a popular agent of rejuvenation. The poet W. B. Yeats tried it in 1934, five years before his death, and was impressed by its results (he became known to locals as the 'gland old man' as a result). The castrated do live longer than the unmutilated and Matthew speaks of 'eunuchs who have made themselves eunuchs for the Kingdom of Heaven's sake' to increase their chances of eternal life. We do not know if they succeeded.

The physician Alex Comfort was among the first to show that hunger improves patterns of survival in mice (although he is perhaps better known for his book *The Joy of Sex*). Comfort finished his career in an attempt to improve the erotic universe of older people. He did not find the task easy, for they – and he – had by then been overtaken by physical decay. His failure to bring joy to the aged is a stark reminder that to indulge in sex levies a penalty in the currency of age. It hints both at why babies grow old and how they are born young.

Alex Comfort himself died at eighty of a stroke, but for most of his clients the Grim Reaper arrived in the form of cancer, which kills more people than does any other disease. Colon cancer, for example, is a thousand times more common in eighty-year-olds than in teenagers. Almost all those over seventy have some form of the illness. Tissues that copy themselves all the time – white blood cells or those in the skin – are more liable to run out of control than those, such as the heart, whose cells do not often divide. Cancer is immortality gone wild, the survival of cells that should die.

Age may be in part a side-effect of genetic mechanisms that protect against such uncontrolled growth. They lose their power as the years go on but as the genes have by then been passed on to the next generation, natural selection does not notice.

Healthy cells kept in culture have a series of internal counters that tell them how many times they have divided and when they should give up. Forty or so rounds of division is the normal limit. Because cancer cells refuse to listen to their instructions they become, in effect, immortal. Part of the complex machinery of cell suicide resides at the tips of chromosomes. Each bears a length of DNA called a telomere, its six-letter message repeated many times. It acts rather like the crimp on the ends of a bootlace and protects the main body of the chromosome from wear. At each division the telomere loses some of its repeats. When it reaches a critical level the cell can no longer divide. An enzyme called telomerase can, given the chance, help to repair the damage, but is more or less switched off in most tissues.

Most people are born with around ten thousand DNA letters in each telomere, but some have twice as many as others. The number is reduced with each division, at a great rate as a child, slower in young adulthood and faster again as the years roll on. In old age, the chromosomal tips may be no more than half the length they were at birth. Wild birds born with short telomeres die rather young, as do laboratory mice engineered to have the same predicament. Another hint of their importance comes from the rare inherited conditions in which young children age far faster than normal. In the most severe forms they die in their teens and look like people seventy years older than themselves. Their telomeres are much shorter than average.

Certain diseases of the elderly such as heart failure are also accompanied by reduced versions of the protective crimp and the shorter the structure the earlier the symptoms. Telomeres are but one step in the process of decay, for as they wear away the mitochondria begin to fail which, in turn, leads to muscle weakness, anaemia and more.

Ageing at the cellular level is complex indeed. However intricate its workings, sperm and egg reset the biological clock, and there the relevant enzyme reveals itself as at least part of the fountain of youth. Telomerase uses its magic to restore chromosomes to full length. It works at full power in testis and ovary and, so refreshed, male and female sex cells meet, and a child is born young. In time it too will age and die – but by then the clock in testis or ovary will be set back to zero for the next generation.

The inbuilt timers show that in the end the body will fail however well its owner cares for it. We die for evolutionary reasons, some of which we understand; and we may hence be able to make some guesses about our prospects.

Humans live longer than do their relatives, with wild chimps lasting around twenty-four years and gorillas five years less than that, but their overall patterns of survival – in which males die younger than females and decay speeds up with age – are not much different from those of other primates. That hints that the arrangement has descended from remote shared ancestors. In contrast, the overall length of life of various species of ape does not fit at all with their biological relatedness. They have not inherited some ancestral scheme of ageing; instead each species – our own included – has evolved its own strategy within the relatively recent past. Even baboons or chimpanzees from different places have different patterns of mortality as a hint that the process is in evolutionary terms quite flexible.

Given the great change in human ecology in recent times, will Darwin's machine change the future of old age, as it has changed the past? The shifts in human mortality from place to place and across the centuries are an experiment in natural selection. That process allowed light skin to evolve from black in a few millennia and the ability to drink milk when adult in an even shorter period. Might the dramatic increase in survival over the past several centuries have led to an equivalent shift in the rate of decline? Most people no longer face the random accidents of a dangerous world but spend their days in a safer and more uniform environment than ever before. Does the mortality clock slow down when life is easy? The answer to that question does not bring much comfort to those who are looking forward to their centennial telegram.

John Graunt had successors across the world, the most efficient among them in Scandinavia. Sweden now ranks high in the survival stakes, at number seven worldwide, with an average life expectancy of eighty-one years. A fifth of all Swedes are over sixty-five. All this is new and for much of the country's history the struggle to survive was even tougher than it was in the British Isles.

The nation has undergone long periods of war and famine, the latter not helped by its position on the northern edge of the habitable globe. In the icy winter of 1696 a third of the inhabitants of some provinces starved to death while endless quarrels between dynasties killed thousands more. Plague, leprosy, smallpox and consumption did even more damage than they had in Britain. In the 1860s no more than three Swedes celebrated their centennial each year, a figure two hundred times smaller than that today. For much of the eighteenth and nineteenth centuries mass emigration to the United States was all that staved off disaster.

Such huge changes in conditions, matched with the nation's excellent records, can be used to ask whether the physical rate of decay has altered as a result of the move from chaos to calm and from poverty to affluence, or whether its progress is due to no more than a fairer society, improved medical care and the like.

In fact, when the statisticians remove the effects of disease, starvation and war, Sweden's death rates – and hence the rate at which its people age – have not changed at all from its turbulent past to its social-democratic present, over more than a dozen generations (which is enough to shift the patterns of mortality in experimental populations of mice and fruit-flies). Its inhabitants are, it seems, close to the physical limits of human longevity, so that social change and medical progress may not take them much further.

In years to come, the Grim Reaper may be pushed back in poor countries as conditions improve, but his scythe swings across Scandinavia and the rest of the advanced world just as hard as it ever has. The real nemesis – the enemy within, the inevitable decline of the body's machinery – has plodded on at the same rate even as our external foes have been defeated. Old age has not slowed down, even if it has a larger, and better treated, constituency upon which to work.

The biblical message is full of reminders that death will not long be delayed, and it is right. Even so, those depressed by the imminence of decline can at least take the comfort offered by God to Job that 'Thou shalt come to thy grave in a full age, like as a shock of corn cometh in his season.' For the first time in history, that solace is available to almost all of us.

CHAPTER FIVE

NO THOUGHT FOR THE MORROW

William Blake, *The Rock Sculptured with the
Recovery of the Ark*

*And thou shalt come into the ark, thou, and thy sons, and
thy wife, and thy sons' wives with thee.*

GENESIS 6:18

I once appeared in a television commercial for an insurance
company. The Equitable Life Assurance group had the idea of
touting for custom with more (or in my case less) well-known
figures happy to expound on the need for cover during their

203

declining years, extended as they have become. My co-stars included the boxer Muhammad Ali and Buzz Aldrin, the second man to walk on the Moon, and the only person to receive Holy Communion on its surface.

The ad seemed to work and thousands shelled out for a policy. They were subscribing to a scheme that diluted individual risk through the population as a whole. Such arrangements bind people into a common intent to share the burden of potential disaster.

Collective purpose of this kind has a lot in common with religion. The Book of Proverbs emphasises the need to plan ahead against difficult times to come with the tale of the ants: 'A people not strong, yet they prepare their meat in the summer', while the New Testament goes further, for it promises eternal life for those who buy in to its doctrines. A millennium earlier, the Babylonian Code of Hammurabi allowed farmers to cancel their debts should they be overwhelmed by a sudden flood. In that agreement, and in its many successors, each side makes its own estimate of risk. Vicars and insurance salesmen are in the same business, for each guarantees support against unwelcome events in times to come to those who invest in their product. They each depend for financial or moral success on the assumption that they are better informed about the costs and benefits involved than are their clients.

A judicious appeal to higher powers has often prevented catastrophe. Gibbon tells of a pre-emptive strike against the stormy sea in which 'Epidaurus must have been overwhelmed, had not the prudent citizens placed St. Hilarion, an

Egyptian monk, on the beach. He made the sign of the Cross; the mountain-wave stopped, bowed, and returned.' In this post-miraculous world, floods give rise to more of a headache. They are responsible for a third of the global costs of natural disasters and kill almost half the victims of such events. To protect themselves, those unsure of the power of the saints can invest in dykes, sandbags and the like. In addition, many buy policies that promise to pay up if the waters rise.

Disasters, whatever their cause, are bound to happen. The difficulty lies in knowing when. Nobody has ever made a successful prediction of an earthquake. Floods, volcanoes and hurricanes give more notice of the possibility of doom, but even they are hard to foretell more than a few days ahead.

Archbishop Ussher's biblical timetable tells us that the first of all flood alerts, and the first insurance policy, came in 2348 BC. In that year, a certain patriarch was told to prepare himself. As God put it: 'And, behold, I, even I, do bring a flood of waters upon the earth, to destroy all flesh, wherein is the breath of life, from under heaven; and every thing that is in the earth shall die. But with thee will I establish my covenant.' Noah and his shipmates survived while all other creatures – men, cattle, fowls and more – were destroyed. He had, like the clients of Equitable Life, entered into a covenant, a contract, with his guarantor. Each signatory played his part in the bargain. Noah, unlike his feckless fellows, was seen as a good bet in the eyes of the Lord and quite soon, his policy paid off.

The insurance industry's biggest problem remains that of the biblical hero. In the past decade it has paid out two hundred billion dollars to those overwhelmed by the waters. In these islands alone the exceptionally wet year of 2007 cost shareholders three billion pounds. Five million Britons will face the risk of inundation by mid-century and the annual pay-out may increase by more than half. A rising river calls for forethought. Households at risk are told to stockpile sandbags, to listen for sirens and to ensure that their valuables are kept on an upper floor. In many places they can be ordered to evacuate if the danger is imminent. If they do not take such precautions they may find it impossible to obtain redress.

The devout sometimes argue that to bet against the future in this way shows a lack of respect for divine omnipotence and that Christians should be advised instead that 'the morrow shall take thought for the things of itself'. Now and again, the insurance industry uses the same Jesuitical logic. Until a few years ago an underwriter could state after a major calamity that his company was not obliged to fulfil his bargain on the grounds that the event was an 'Act of God': an unpredictable catastrophe that could not be built into the normal perception of risk. In these secular times the language has changed, but the argument has not, for after a huge natural disaster those expected to pay out may still slip gracefully from their contract with an appeal to the power of Nature and, in some places, to the state. As science improves our ability to work out the chances of fire or flood that excuse is applied less than it was, but customers can still be outraged by the secular priesthood's refusal to live up to what their flock, with their regular contributions, imagine to be its responsibility.

Acts of God are a two-edged sword. In 1915 a huckster who claimed to be able to summon up rainstorms with a chemical mix was hired by the city of San Diego to break a drought. It then rained mightily, but his success came at a price, for the deluge washed away a hundred bridges and killed twenty people. The city fathers withheld his fee on the grounds that the rainmaker was liable for the damage. He counter-claimed that the inundation was a result of divine providence, but that invalidated his contract and he walked away penniless.

The 1953 Essex Flood was the biggest British natural disaster of the twentieth century. Three hundred people drowned (and the Dutch death toll was six times greater). A powerful basin of low pressure moved down the North Sea. It whipped up a storm surge with tides six feet above normal, topped by thirty-foot waves. The dykes had been neglected during the war and the sea smashed through in a thousand places. Water deeper than the height of a man raced through Harwich. No alarm system was in place and many died because the flimsy prefabs and caravans in which they were forced to live during those hungry times were washed away. More than twenty thousand homes were damaged or destroyed, some as far up the Thames as the East End of London.

Memories of such events echo through the years. In 1725 a strange relic was found in a quarry near Lake Constance. It was hailed as *Homo diluvii testis*; 'one of the rarest remains of the accursed race buried beneath the waters … The dismal skeleton of an ancient sinner, drowned thusly in the Deluge.' The bones were taken as proof of the biblical Flood. Their discoverer, Johann Jakob Scheuchzer, knew just what had happened. The hand of God had stopped the Earth's rotation, and the seas – the 'fountains of the deep' – had flowed over the land. He exhorted those sceptical about the Genesis account to 'come and watch ... and the dumb rocks will preach it to you'. Many did and were convinced (although Voltaire argued instead that the preserved fish found in the Alps were no more than the remains of snacks dropped by Crusaders). In 1811 the anatomist Georges Cuvier showed Diluvian Man to be a petrified salamander rather than a drowned sinner and soon afterwards the idea that rivers, seas and landscapes themselves could rise, fall and reinvent themselves became a central part of geology.

The Ark, the Deluge, and the shipwreck on Ararat tell the tale of a great cataclysm. They remind the faithful of the power of God to move the Earth as a punishment – or a warning – to sinful Man. Again and again, the deity shakes his spear: at the fall of Babylon as viewed in the Book of Revelation 'there were voices, and thunders, and lightnings; and there was a great earthquake, such as was not since men were upon the earth, so mighty an earthquake, and so great'. Sodom, too, got it in the neck for its filthy ways when brimstone and fire were rained upon it, while in Revelation the signs of God's wrath are further manifest in the form of roaring waves and 'a great hail out of heaven, every stone about the weight of a talent'. Jesus himself warns that at the end of time there will be 'famines, and pestilences, and earthquakes, in divers places'.

The notion of disaster as divine reprimand is hard to resist.

In 1750 John Wesley in his 'Sermon on the Cause and Cure of Earthquakes' argued that such tremors began only when the Serpent ruined it for everybody; 'before the sin of Adam there were no agitations within the bowels of the earth, no violent convulsions, no concussions of the earth, no earthquakes, but all was unmoved as the pillars of heaven'. He was supported in 2007 by the Bishop of Carlisle who claimed that the severe floods in his diocese were due to godly disapproval of British plans for gay marriage (and Hurricane Sandy, which devastated New York five years later, was also said to be sparked off by sodomy and to be a sure sign that the End Times are near).

Whatever its cause, the Genesis Flood has a basis in history for in the Bible Lands and elsewhere such events have taken place again and again. Those who take thought for the morrow fear that, in this era of climate change, a similar disaster will soon be upon us and that it may already be too late to protect ourselves against the deluge to come.

Geologists after Darwin were steeped in the notion that landscapes were built by gradual processes that can be seen at their patient work today. Their world has moved on. Sudden catastrophes such as earthquakes, eruptions and floods are now at the heart of their science.

Most cultures have folk tales of terrible disasters in ancient times. In the fourth century BC, the Greek philosopher Euhemerus came up with an explanation of what lay behind them. Myths of divine displeasure were not evidence of supernatural intervention but were embellished accounts of real occurrences. He was right, and his idea has become a profession. Geomythology ties legend to science and asks how, why, where and when the incidents recalled took place. Ancient fable and modern technology each have lessons for the future.

Every agent of inundation has a match in legend. The Psalms tell how an earthquake disturbed the waters: 'Jordan was driven back. The mountains skipped like rams and the little hills like lambs. What ails thee, O thou sea, that thou fleddest? Thou Jordan that was driven back?' The fall of the walls of Jericho may also have been due to a similar event, with Joshua's trumpet blast as its grumbles. In 1927 another incident of this kind damaged the Church of the Holy Sepulchre in Jerusalem.

The fire-dragon known as the Chimaera was, legend tells us, slain at her lair by the hero Bellerophon. Her fiery breath lives on. She can be visited on the Turkish coast, where a jet of methane from underground has been alight for millennia. Not far from her nostrils lie the ruins of the city of Colossus. In AD 60 an earthquake struck. The town was built over a rift in the Earth, where a malodorous spring rose from Hades and emerged into the temple (the Delphic Oracle lived in a similar spot and gave her finest prophecies after a few breaths of its gases). The quake was explained by the Greeks as a visitation of the snake goddess Echidna, but as Christianity – with the help of a series of epistles from St Paul – gained hold, the tale grew that the Archangel Michael, patron saint of grocers, was responsible instead. His presence shook the ground and opened a wide canyon as he raised thunderous voice against heresy. Loch Ness, too, lies over the Great Glen Fault and the first record of the Monster, in the seventh-century *Life of Columba*, tells of 'great shakings' when it appeared.

Jolts of this kind beneath the waves can be lethal. The surge that followed the Lisbon Earthquake of 1755 drowned thousands, many of them far from the upheaval itself. It washed people away in Barbados, and destroyed Galway's city wall. The explosion of Krakatoa was heard in Sydney and its waves even reached Dover.

The first tsunami to be recorded came in around 1620 BC with the explosion of Thera. The eruption – its remains now the island of Santorini – was more powerful than that of Krakatoa and generated a wall of water that swept across the Mediterranean. It was once thought to be responsible for the extinction of the Minoans on Crete but now it seems that an earlier burst of ash gave them time to flee. The Egyptians spoke of fire and floods, the Greeks – upwind of the explosion – of a noisy battle as the gods threw great stones, while in Turkey the Hittites, who saw no more than smoke, remembered a giant who touched the sky.

More floods are sure to come. Seattle seems a phlegmatic sort of place, but beneath the stolid soles of its citizens the town is skittish indeed. It sits on the youngest and hottest active zone in the world. Far below lie faults that strain against each other, ready to slip. Out at sea rock is dragged into the depths and over-ridden by another continental plate in its journey across the globe. The place is poised on the edge of chaos. It faces a one in twenty chance that within the next decade there will be an upheaval worse than the 1906 San Francisco event. Should the slip be in the closest fault, a fifteen foot wave will reach the shore within twenty minutes, leaving almost no time for escape. Further up the inlets that seam the coasts the waters would rise ten times higher as the creeks narrow.

Long before the city was founded Native Americans lived nearby. The descendants of the Duwamish people tell tales of a fearsome snake with antlers, an *a'yahos*, a creature best avoided because it caused the ground to shake and all who saw it to turn to stone. The record of the rocks shows that the land upon which their ancestors lived did indeed move and must have killed many. The last great quake, in around AD 900, caused the surface to slump by fifteen feet. It drowned vast numbers of trees (whose trunks still remain beneath the waves) and laid down sheets of sand high above the modern shorelines. A series of such ghost forests, of ancient landslips, and of cores through the soil shows there have been seven great earthquakes in the past three and a half thousand years. The 'spirit stones' into which their unfortunate witnesses were transformed are the remnants of landslips that date from the most recent event, which has lived on in collective memory for more than a thousand years.

Tribes further down the coast have no fear of an *a'yahos*, but are instead concerned about gigantic versions of real animals. Their legends speak of a struggle between Thunder-bird and Whale. The bird, a huge raven, bore a lake on its back and the mammal ruled the seas. The raven attacked the whale, and was dragged into the depths. The commotion caused the sea to overflow. The evidence of a tsunami that struck three centuries ago is preserved as a thick layer of sand that sits on top of the peat of the local marshlands.

That tale has an echo far away. In Japanese legend, tsunamis are caused by the uneasy movement of a giant catfish, held down most of the time by a god who now and again has to leave for a conference with his fellow-deities. The Edo dynasty kept sophisticated records, and on January 26th 1700 they tell of a frightful wave, which without doubt emanated from that self-same disturbance in the Pacific north-west, the outcome not of a restless fish, or a battle between raven and whale, but of an undersea tremor.

Catfish, whales and thunderbirds have been superseded by strain gauges that scan the ocean floor, together with hundreds of buoys used to sense the passage of a watery mountain. The coming deluge will be hailed by sirens and speakers scattered through Seattle, Tokyo and other coastal cities. In spite of the technology, nobody has yet managed to forecast a submarine heave, although satellites can now pick up the great wave as soon as it begins to move.

The Boxing Day tsunami of 2004 began off the coast of Indonesia as a rupture between the Burmese and the Indian continental plates. A line of rock fifteen hundred kilometres long jumped forward by twenty metres. At once, the sea floor on the Burma side rose. It displaced a vast mass of water that travelled at five hundred kilometres an hour across the Indian Ocean. The flood killed more than two hundred thousand. The 2011 event in Japan was sparked off in the same way and the slip was so close to the coast that there was almost no time for the alarms to sound. It drowned ten thousand.

Such events have also happened closer to home. In around 6000 BC an underwater landslide off the coast of Norway set off a great wave. The water hit the British coast with such force that it caused the final breach between Great Britain and the Continent. The breakers forced themselves twenty miles inland, far further than in the Burma and Japan tsunamis. They must have overwhelmed many of the scattered hunters who lived there. Water can also be displaced when sections of unstable volcanic ash slump into the sea. In the Pacific seven once-inhabited islands have disappeared within historical times because their submarine slopes collapsed, taking the land and those who lived upon it with them. Fragments of coral splashed a thousand feet high on the Hawaiian Islands show how huge the surge from such upheavals may be. An unstable undersea structure of this kind sits off the coast of Tenerife, and should it slip a wall of water the height of an office block would strike southern England within three hours.

Disaster from above can cause as much damage as that from below. Maori and Aboriginal myths tell of falling stars, of fire gods and of deluges from the sky. Five hundred years ago, a great wave struck New South Wales and the South Island of New Zealand. It roared over the coastal headlands, some of which still bear the remains of ancient camps, their inhabitants washed away. It was generated by a comet that struck the globe. Its crater, twenty kilometres across, is two hundred kilometres south of New Zealand. The impact was equivalent to an earthquake bigger than that of San Francisco and generated a watery mountain that reached far inland.

For anyone who lives close to the sea or to a great river catastrophe is, in the end, inevitable. The largest floods are transformed into myths, but the memory of many more – and of others greater yet – lives on in the records of the scribes and of the rocks.

The Egyptians held that the Nile rose each autumn because of Isis' tears for her husband Osiris. They measured its flow for five thousand years. The figures were kept – with occasional gaps due to political upheaval, to shifts in the river, and to overt falsification – to plan the need to guard against starvation with sufficient stores of grain and to work out the tax rate; high in good years, low in bad. The marked sticks first used to make the measurements evolved into gauges with channels that led from the river into a graduated cistern, often within the grounds of a temple. Most have been lost but a survivor, the Nileometer, still stands near Cairo, not far from where Moses was found in the bulrushes. Today's version was built after the Arab conquest in the ninth century but it succeeds a structure that had been there since pharaonic times. Many such stations were scattered along the river. Fast rowers sped from far upstream to warn how high the water would rise and what that would do to the crops. Pliny named six levels of alertness; starvation, scarcity, cheerfulness, safety, exultation, and – now and again – destruction.

The Nile records are the longest-running set of scientific data ever gathered. They hint at a basis for another biblical disaster. The Nile's rise and fall depends on rain in the highlands from whence it springs. Its rhythm arises from a regular shift of the patterns of high and low atmospheric pressure in the North Atlantic. A statistical filter applied to the figures from AD 622 to 1922 points at a seven year cycle of drought followed by another seven of steady flow. Genesis tells us that the Pharaoh was baffled by his repeated dream of seven lean and seven fat cows and that only Joseph could interpret it: 'Behold, there come seven years of great plenty throughout all the land of Egypt: And there shall arise after them seven years of famine; and all the plenty shall be forgotten in the land of Egypt; and the famine shall consume the land.'

In an early example of insurance advice Joseph counselled the ruler to store up a fifth of the harvest in the years of plenty as a reserve against the shortages to come. A deeper analysis of the Nile records hints at longer patterns of around ninety and two hundred years. They might result from changes in the output of the Sun.

Lesser watercourses can also cause plenty of damage when they burst their banks. As the snows melted after the bitter winter of 1947, the Thames at Maidenhead was a mile wide. In 1928 fourteen people drowned in their basements when the Chelsea Embankment collapsed. That crisis was eclipsed by an 1894 event in which a third of the average annual rainfall fell in a week, and ninety years earlier *The Gentleman's Magazine* recorded that 'In the neighbourhood of Kenning and Vauxhall, a torrent of water has arisen, which in its progress has carried away furniture, trunks of trees, cattle etc., and has destroyed a great number of bridges .... The flood has been very general, and tremendously fatal in various parts of the country.' At Oxford in 1593 matters were even worse, for Christ Church Meadow was beneath thirteen feet of water while in 1236 'in the great Palace of Westminster men did row with wherries in the midst of the Hall'. A century and a half earlier, London Bridge was swept away, with many deaths on the bridge itself and upon the river's crowded banks. A layer of river gravel on the local plain hints at a huge incursion four thousand years ago that dwarfed all those events.

That inundation, the Nile floods and the Boxing Day tsunami, were in historical context trivial. The watery legends of the Greeks are grander than those of Native Americans. They had, they imagined, roots in ancient times. The Ogygian deluge is said to have taken place at the time of the first rulers of Attica, which Plato dated at ten thousand years before his day.

Should the event indeed reach back to that distant age it would coincide, more or less, with the end of the last glacial period. A great surge then rocked what was to become the cradle of civilisation. Its scenery, like that of other coastal states, was transformed as the sea rose by a hundred metres. A landscape of hills became a pattern of islands. Until the waters came, the picturesque patchwork of the Cyclades – Mykonos, Naxos, Thera and the rest – formed a land mass called Aegis and the Gulf of Corinth was a dry valley. The preservation in myth of an event so remote might be hard to accept, but Australian dream-time stories tell of an era when what is now a great bay in South Australia was itself open land. Geology shows that it was submerged ten thousand years ago. The persistence of memory is such that the locals can even recall the names of the places drowned at that time.

There have been at least five major eons of ice since the Earth began, and within each there have been many expansions and contractions of the cold shroud. Altogether they account for no more than a minor portion of our planet's history, but some episodes lasted for more than two hundred million years and for a short time ice may have covered the entire globe. The persistence of the Arctic and Antarctic sheets means that we are ourselves still in such an era. It began around two and a half million years ago and has ticked back and forth between glacial and interglacial periods with a rhythm of between forty and a hundred thousand years since then.

The agents that drive the glaciers work on many scales. Over years to decades, shifts in snowfall or sea ice, or cloud, or dust in the air can lead to change. As decades stretch to centuries, the Sun's output rises and falls and the Earth's orbit alters, while the upper layers of the ocean warm and cool to give events such as the 'Little Ice Age' that stretched from the sixteenth to the nineteenth century. As centuries give way to millennia the growth or loss of forests makes a difference and sea water from the depths to the surface may lose or soak up energy. On that scale ice sheets grow and shrink (and reflect more, or less, solar energy back into space), and methane compounds hidden beneath the oceans release the gas or soak it up. Over geological time disturbances such as continental drift and the greenhouse gases that escape from volcanoes and huge fires make a difference, as do chemical shifts in air and water as rocks are worn away or are sucked into the depths. All these variables interact in complex and unpredictable ways.

Whatever lies behind them, glacial periods tend to arrive slowly but to go out with a bang. The most recent freeze began just over a hundred thousand years ago, with a gradual decrease in temperature that lasted for tens of millennia. At its peak, a cap of frozen water covered much of North America and Northern Europe (most of the British Isles included). Greece had glaciers of its own and the seas around Attica were as frigid as today's North Sea. In some places the pallid shroud was three kilometres thick, and so much water was trapped that global sea-levels were more than a hundred metres lower than those of today.

Before its final collapse that harsh and icy era was interrupted by several brief warmings which lasted for a few thousand years before the return of the chill. As glaciers met the seas great chunks of ice broke off. Fleets of icebergs, filled with detritus smashed from the rocks below, set off into the oceans and, as they melted, shed their load into the depths. Today's mid-Atlantic floor bears rocks from Greenland, Europe and North America as evidence of their passage. Again and again, the ice recovered and the winter went on.

Then came the end. The climate reached a tipping point. The evidence lies in ocean mud, in fossil pollen, and in the ratio of chemical isotopes in stalactites and other sedimentary rocks that change with shifts in temperature. Around fourteen and a half thousand years ago, the climate began to switch from cold to warm and back again until, one day, a warm period ran away with itself. The Earth reached a new equilibrium. The biggest shift was packed in to little more than fifteen hundred years from around thirteen thousand years ago. Almost at once, the white mantle was gone and the Ogygian inundation was under way.

The polar ice sent out a vast – and final – armada which covered much of the northern seas. A slight increase in the Sun's output was matched by the disruption of deep ocean currents caused by cold water as it sank from above. The Age of Ice was close to its end. In its decline it fed upon itself and in its death throes drowned the world.

The deep seas are a vast reservoir of carbon dioxide, dissolved under pressure, but the cold and weighty water sank to the bottom and pushed that ancient store towards the surface. The pressure dropped and, in a subdued version of the fizz of a bottle of champagne, greenhouse gases escaped and entered the air. They pushed onwards the wave of warmth.

As the glaciers fell back water streamed off the land. By 5000 BC all that was left of the North American ice sheet was a small cap on Baffin Island. Its Old World equivalent retreated to the North Cape. Greenland alone kept its cover. Sea level rose by a hundred metres and more (a solid portion of which was added in just a couple of centuries as ocean currents brought a pulse of tropical warm water to northern shores). With a few interruptions the Earth began to experience the climate of today. Life entered a new phase. Vast numbers of large animals – American ground sloths, woolly mammoths, kangaroos the size of a rhinoceros – disappeared, because they were hunted to extinction by the booming human population and because their homes were destroyed as the weather changed. Many died out within no more than a few centuries.

Together with the Somme, the Rhine, and the Elbe, the Thames became no more than a tributary of a gigantic waterway, the Fleuve Manche. The Fleuve ran down the centre of what is now the English Channel and for a time filled a great lake in the middle of today's North Sea. As England's glaciers retreated, the land crushed beneath their weight rose upwards to speed the torrent. Probes into the sea floor far into the Atlantic reveal great beds of mud, the relics of a destroyed countryside. The flow of the Fleuve was at its peak equivalent to that of today's silt-laden Congo (which is far longer, with a much more extensive basin). As its waters poured into the ocean, they further disrupted circulation in the deeps and hastened the warming of the world. Within a few centuries, the Ice Age was over and the continents gained much the shape they have today.

Even places far from the coast were marked by the great flood. The 'scablands' of the American north-west were named by the pioneers for their inhospitable deep valleys and stark plains of bare and shattered rock seamed by canyons. They stretch for two hundred miles from Spokane in Washington State to Portland in Oregon and were formed in almost an instant, as a monument to the final days of the great American glacier. A half-mile-high dam of ice that held back a vast body of melt-water called Lake Missoula shuddered and broke. Within a day, five hundred cubic miles of water travelled at eighty miles an hour to flood three thousand square miles of eastern Washington. At its peak, the flow was equivalent to ten times that of all the modern world's rivers combined.

Over the next two weeks the deluge drained with enormous force through the narrow Columbia Gorge and headed towards the ocean. Its effects were impressive. They can be seen today as scars on the rock of the gorge a thousand feet above its floor, and as ripples of gravel that look like the patterns on a wave-lapped beach – but not as tiny ridges a few inches apart but mounds of shattered rock thirty-five feet high and several hundred feet from each other. The ice-dam reformed and broke again and again. Native Americans might have seen the Missoula Deluge but, if they did, their legends have been lost.

Soon, more lakes burst their bonds. In the final decades of the North American Ice Sheet came the collapse of a frozen dam that held the Canadian Lake Agassiz (which held as much as all modern bodies of fresh water put together). The water smashed through and roared into Hudson Bay. Samples through peat bogs as far away as Holland show that within two centuries the Dutch coast faced a two metre increase in the height of the tides. So much icy water escaped in that deluge that the world cooled even as the oceans rose.

Might the story of Noah be tied to such an episode? Three hundred Flood narratives are known from across the globe (an Australian version tells of the frog that swallowed the world's water but spewed it out in a great inundation when the other animals managed to make him laugh). Some have a strong echo of the Genesis story. A Babylonian tale speaks of a divine decision to destroy everyone on Earth, apart from a ruler called Atrahasis, who is warned by God to build a boat for his family and animals. He heeds the advice. The waters rise, he floats off, and, in time, drifts to shore. A real Atrahasis ('the very wise') ruled in Sumeria around 3000 BC, in the Early

Bronze Age. Excavations of the remains of his city, in modern Iraq, show signs of a gigantic upheaval of the Euphrates at just that time. Its clay tablets tell of the death of multitudes, with the river so full of corpses that they looked like grasshoppers.

Other candidates for the biblical Flood come from long before the Sumerians. At the end of the Ice Age the Old World had its own Lake Missoula. Twelve thousand years ago, when the cold was in full retreat, the ice-dammed Lake Kuray-Chuja formed in the Altai Mountains of southern Siberia as engorged rivers were trapped by a glacial tongue. The lake was half a mile deep and held five hundred cubic miles of water. The barrier collapsed and the mass surged downstream. Like Missoula, Kuray-Chuja emptied within a single day. The flood roared west and south into the Aral Sea, which overflowed, and then into the Caspian, which may have spilled over into the Black Sea, not far from where Mount Ararat stands. In time its contents drained into the Mediterranean.

The shores of the Black Sea do show evidence of rapid changes in water-level at around that time and later. A hundred metres down lies an ancient shoreline of around 5000 BC that might mark the sudden inundation of as much as a hundred and fifty thousand square kilometres of land. Whether such events can be blamed on floods from Siberia is not clear. The Mediterranean would be a closer source. It was once claimed that, at the end of the Ice Age, salty waters broke through a barrier where the Bosphorus now lies and filled the basin within a few weeks. The sudden inundation, so the story went, drove its inhabitants away and gave rise to the Flood legend. The fossils of water-snails from the Danube delta tell a less dramatic tale. They show that the Sea has been a fresh-water lake for most of its history, with a level that fluctuated around thirty metres lower than the present level. The shells hint that salt water did break through around nine and a half thousand years ago but as an ebb and flow between lake and sea as the oceans rose and fell rather than a deluge. In time the Black Sea became as saline as it is today. Water still pours into it from the Mediterranean, but far below the surface, in an invisible river three hundred times larger than the Thames that rushes past Istanbul. Its waters are salty and heavy and sink to the bottom, where they have carved out a channel half a mile wide, with its own submarine rapids and waterfalls.

Men and women have lived on Black Sea shores for tens of thousands of years. In dry and icy periods, when the Sea shrank, they moved away from its cold soils, but they came back to hunt and then to farm as the weather improved. For them, the risen waters meant not flight, but a settled existence on a wooded and fertile landscape. Even so, the coincidence of the rise and fall of that sea with the story of Noah is, like the legends of the Australian Aboriginals, hard to dismiss.

If the tale of the Flood has some basis in truth, what of the vessel that floated on it? Natural arks are real enough. Rafts of vegetation can drift for thousands of miles. As a result, remote islands that emerge from the sea through volcanic action soon gain a life of their own. Plants and animals that can fly, blow, or swim tend to arrive first, and frogs, which cannot stand salt water, never make the trip. In New Zealand, the molecules show that many plants floated across from Australia – fifteen

hundred kilometres away – not long ago, for they are so similar to their cousins across the Tasman Sea that they cannot
have been in their new home for more than a few millennia.
Chameleons are among the hardiest travellers, for they have
made it from Madagascar to the mainland of Africa at least
three times, to give rise to three distinct African groups of
those reptiles.

The biblical equivalent must have been crowded indeed.
Genesis tells us that all creatures that walk on the ground
found a place, while Leviticus later adds those that 'whatsoever hath more feet', which gives spiders a chance. Fowls were
allowed on, too. A strict interpretation insists that there must
also have been giraffes, kangaroos and more for they exist
today even if they were unknown to the authors of Genesis.
Literalists even squeeze in dinosaurs. More realistic readings of
the text suggest that such exotics were so unimportant in
God's eyes as to have been allowed to survive in some remote
region. Seven pairs of each clean animal were given entry but
the others came just as couples.

Many Arks have been built since Old Testament times. In
the seventeenth century a Dutch version three hundred feet
long was said to be so seaworthy that copies were made, while
in 2007 Greenpeace assembled a replica vessel high on Ararat
to emphasise their concern about sea-level rise. Genesis gives
the original's length as three hundred cubits, which means a
ship five hundred feet long – sixty feet more than the
schooner *Wyoming,* the largest wooden vessel of modern times
and one so unseaworthy that even moderate waves caused the
seams of her hull to open. She was in the end lost with all

hands. A full-scale version, the $150 million Ark Encounter, is under construction in Kentucky, not far from the infamous Creation Museum. It will house animals such as giraffes (young ones, for its builders argue that God used juvenile animals to save space) but will remain on dry land. A floating version of the same size has already been built in Holland, but that is rather a cheat because it rests on the hull of a steel barge (it is, however, blessed with a petting zoo).

Biblical scholars are also concerned with the question of quite where Noah's vessel came ashore. Over two hundred people claim to have seen its remains, but the million dollar prize offered by a Christian foundation for physical evidence of the mythic craft remains unclaimed. Those who search – often obsessively – for it are known as 'Arkeologists'.

They have many predecessors. In 1571 Marco Polo wrote that 'In the heart of Greater Armenia lies a very high mountain, in the shape of a cube, where, some say, Noah's Ark still lies . . . At its top is snow, so much that nobody can climb it.' In the fourth century AD, a St Hagop took the first step towards their science. He set out to climb Ararat's five thousand metre peak in search of Noah's lifeboat but each time he fell asleep on the ascent, he found himself at the foot of the mountain when he awoke. In his dreams, an angel told him that the place was forbidden territory, but as a consolation prize the heavenly messenger gave him a piece of the vessel's wreckage, which can still be seen – along with the lance that pierced Jesus' side – in a local monastery.

In 1829 Nicholas I of Russia attacked the Turks, in part to gain control of the sacred peak. A German explorer, Friedrich

Wilhelm von Parrot, who served in the Tsar's army, set out to conquer Ararat. A band of soldiers and monks carried a large cross almost to the top but bad weather forced them to descend. A few days later, with a much smaller crucifix, they made it to the summit, where in full dress uniform they drank the health of Noah. Fifty years later a British diplomat, James Bryce, found a piece of wood on the icy cone and at once hailed it as a relic of the Ark (it may instead have been a fragment of von Parrot's cross). Others who might have seen the wreck include a group of Cossacks who were sent to the region in 1916 and noted a bowsprit that poked from a lake on the mountain's slopes. They were, some say, shot on the orders of Trotsky.

For the people who live around the mountain itself Noah's mythic tale has become a useful lure for tourists. Local wine-makers insist that their grapes are descendants of those planted by its captain when the waters receded. In some sense they may be right, for the earliest known domestic grape seeds are indeed found nearby. In a salt-mine now used to give relief to asthmatics the patients are told that the dense white rock that surrounds them was laid down when the deluge evaporated. The natives are passionate in their certainty that Noah stepped ashore at just that point.

Genesis does say that his ship 'rested in the seventh month, on the seventeenth day of the month, upon the mountains of Ararat' – but what does that mean? In those days the place-name was a general term that referred to what we now call Armenia, which reached well into modern Turkey, rather than to a specific peak. Another option for the spot where the patriarch 'went forth, and his sons, and his wife, and his sons' wives with him' is a mountain called Cudi (which itself means no more than 'the heights') on the borders of Turkey, Iraq and Syria. At the time of her visit there in 1910 the explorer Gertrude Bell found a stone structure called by the inhabitants the Ship of Noah, and noted that each September local Christians, Jews, Muslims and Yazidis gathered in a joint ceremony on the summit to commemorate the end of the Deluge. Not long ago, another hill, on the slopes of Ararat itself, was renamed Cudi on the orders of the Turkish government, who were anxious to draw tourists away from the border, and that too is now the target of obsessives.

Since the time of Gertrude Bell the vessel's remains have been found again and again. In 1949, a United States military flight spotted an 'Ararat Anomaly' from the air. It turned out to be a patch of discoloured ice. Another candidate is a ship-shaped block at around six thousand feet. In fact this is a large piece of limestone which has fallen off a nearby cliff and has, over thousands of years, slid downhill over a clay soil as it froze and melted, helped on its way by the occasional earth tremor.

In 1971 Jim Irwin followed in the footsteps of Buzz Aldrin, my fellow participant in the Equitable Life advertisement, to become the eighth man on the Moon. There he picked up a stone which, because of its extreme age, was named the Genesis Rock. He took a copy to Ararat in an expedition to find the Ark. As he said: 'I sincerely believe that God will allow me to find something on Earth from the Book of Genesis that is more important than the Genesis Rock.' He died of a heart attack before he succeeded in his quest.

Geologists are bored and irritated by the Arkeologists for Ararat has a remarkable history of its own. It sits on the edge of two continental plates in an active volcanic zone. The mountain emerged as a mound of ash and lava rather like Vesuvius, with its last continuous eruption twenty thousand years ago. It wakes up from time to time and patches of new lava point to a series of recent emissions. In a settlement upon its slopes overwhelmed four and a half thousand years ago, a mass of pots and bones shows that the locals did not escape. Legend hints that there have been several great upheavals since then. The last major event, in 1840, set off a landslide that shaped the present summit and caused thousands of deaths.

In 1988 the geological plate upon which the mountain stands lurched forwards once more. The Armenian earthquake of that year killed tens of thousands.

The toll might have been far higher. In 1970 the Soviets had built a nuclear power station at Metsamor, near the epicentre of the tremor. Its reactors were shut down after the quake, but reopened in 1993. The station is now among the last in the world not to have a containment barrier.

Metsamor was built, along with hundreds of its fellows across the globe, because of the fear that oil and coal will, sooner or later, run out. That fear – in the long term no doubt justified – is matched by concern that the reckless use of fossil fuels might lead to a new Flood; to an event like that at the end of the Ice Age which, for much the same reason, will change the shape of the world. Admittedly, God does say to Noah after the Ark has landed that 'I will establish my covenant with you; neither shall all flesh be cut off any more by the waters of a flood; neither shall there any more be a flood to destroy the earth', but his promise looks increasingly hollow.

The idea of a coming deluge caused by a global rise in temperature attracts both support and scepticism. An eminent Arkeologist, a Dr Allen Roberts, armed with a degree in Christian Education from Florida's Freedom University, once sought funds in Australia to excavate the possible remains of Noah's lifeboat. A Melbourne geologist, Ian Plimer, decided to take action. He asserted that Roberts had misled his backers and was, as a result, in contravention of Australia's Fair Trading Act. He took Roberts to court. Plimer lost his case, for the Bench decided that the issue was not within the bounds of Australian civil law, and he was forced to pay hundreds of thousands of dollars in costs. In his book *Telling Lies for God*, Plimer launched a passionate attack on enthusiasts for the literal truth of the Bible, floods and all.

Several years later the *Sydney Morning Herald* lamented that the geologist was still the target of 'fundamentalist organizations ... true believers who claim that they alone have the right to speak'. By then, though, many of his opponents were scientists, for Plimer had become a fervent opponent not just of Arks but of the whole idea of man-made climate change. Much of the increase in carbon dioxide production seen over past decades comes, he says, not from industry, but from volcanoes (the United States Geological Survey disagrees). Plimer derides the notion of man-made global warming as a threat to the future. Millions of people agree with him, but as the years roll on the evidence has become hard to reject.

Many of the deniers are deluded or, worse, in the pay of the oil companies and their shadowy supporters.

The recent shift in climate comes from our attempts to get back to the tropical Garden of Eden where we evolved. To do so we use central heating, air conditioners, flights to the sun and factories that churn out gases as they produce the goods that make life easy. The population explosion and the new hunger for meat add to the problem. Almost half the Earth's surface is now given up to farms or to towns. That ecological shift – which has taken only a few centuries – marks a bigger assault on the landscape than did the end of the last glaciation, when no more than a third of the globe was abandoned by the ice. Cities use vast amounts of energy, but farms are almost as bad. Half the protein in the body of every non-vegetarian owes its existence to a high-temperature (and high-carbon) technology which links nitrogen and hydrogen to make ammonia, and from that to produce the artificial fertilisers now central to agriculture. Half a billion tons of the stuff pours out every year and without it two billion people would starve – but most of what is spread on the fields is washed away or escapes into the atmosphere as nitrous oxide gas, where it joins the tons of methane that escape from the rear ends of pigs and cows. Each is a greenhouse gas more potent than carbon dioxide.

At periods in the distant past levels of that gas were much higher than they are today, while at others they were lower. Changes in average temperature and in its concentration go together over geological time to provide a strong hint of a causal link. A survey of the shifts in climate over the past sixty-five million years in relation to its concentration shows that a doubling of the amount in the atmosphere leads to an increase in temperature of between two and five degrees Celsius.

Sceptics claim – and they are right – that there have been many episodes of warmth in the past that emerged without human help. They insist that the latest is no different. There they are wrong. A 2008 survey of thousands of Earth scientists found that the vast majority agreed that temperatures have risen since 1800 and that almost as many consider that human activity is involved. Experts suggest that the total output of carbon dioxide must stay below a thousand billion tons in the first half of the present century if the world is to have a better than evens chance of limiting the temperature rise to two degrees. On present trends, that barrier will be breached in little more than a decade.

Truth is not defined by opinion polls but the consensus is clear. Even so, a quarter of all Britons deny that the climate has changed at all, as a reminder that scientists do not always get their message across when a well-funded, dishonest and secretive opponent stands in their way.

One great shift came fifty-six million years ago when, in what geologists view as a short period, temperatures rose by five degrees. The effects were dramatic. Tropical creatures such as turtles moved towards the poles and the oceans lost vast quantities of microscopic life as the water warmed and became more acid. The Earth's crust was in turmoil as the shards of what had been the great continent of Pangaea continued their collapse. Volcanoes and earthquakes rocked the landscape.

They were followed by huge fires of coal, oil and peat. Carbon dioxide poured into the air, and as the waters warmed reserves of methane held in semi-solid form melted and joined the other greenhouse gases. The temperature shot up. A core through Spitzbergen shows the rate at which carbon was added to the atmosphere. For twenty thousand years, two billion tons of the element was added each year. That amount, impressive as its effects were, is just a tenth of what is pumped out every twelve months today. Plants and animals are again on the move and the oceans have become more acid as the world relives that ancient crisis of long ago.

The worst is yet to come. For half a million years before the Industrial Revolution the level of carbon dioxide in the atmosphere fluctuated between 180 and 300 parts per million. Since that economic earthquake it has shot up. In the years from 1958 (when direct measurements began) to today, the concentration has gone from 315 to 400 parts per million. In spite of controversy as to how much natural feedbacks caused by melting ice, rising seas and dying plants will increase or reduce the direct effect of the gas, experts predict a period of increased temperature. 2010 was one of the three hottest years since figures were first collected and 11 of the 12 hottest years on record have occurred since 2001. Since 1800 global temperatures have risen by almost a degree Celsius, with most of the increase in the past forty years. We are, as a result, on the edge of an era in which the mercury will rise higher than at any time since modern humans evolved. Inevitably, the ice will shrink and the seas will rise to match.

Many Alpine glaciers have pulled back by more than a

kilometre in the past century. In the hot summer of 2003 a Swiss glacier withdrew far enough to expose clothes made of fur and leather that dated from four and a half thousand years ago; and as these materials break down when exposed to the air they must have been frozen for all that time, as proof that the river of ice is now smaller than it has been for almost five millennia. Information on a hundred thousand such structures across the globe suggests that they will soon lose a quarter of their volume. The Alps will be stripped of three quarters of their ice and New Zealand of almost as much (although the Himalayas have been more resistant to the great melt).

The waters of Missoula burst into an empty landscape, but nowadays great cities teem downstream from the great mountain ranges. In 1892 the collapse of an ice dam released two hundred thousand cubic metres of water from the Glacier de Tête Rousse in the Alps. It killed two hundred in the small town of Saint-Gervais-les-Bains downstream. Around fifty such structures have now formed in the Himalayas. Some of the lakes they hold are a mile long and pose a real threat to people further down the valley.

The poles give even more cause for concern. In September 2012 the Arctic ice fell to a record low of three and a half thousand square kilometres – less than half its extent forty years earlier. At that rate it might disappear altogether in summer a decade from now, opening up the North-West Passage for much more of the year. As that ice floats its loss will make no difference to sea levels but is a warning as to what is happening at the northern and southern ends of the globe. In that same year the greatest heat wave for a century washed over

Greenland, and almost all the surface melted to some degree. The island's summit, three kilometres above sea level, saw liquid water for the first time in more than a century. A global survey based on data from satellite radar and gravity probes since 2005 shows that Greenland lost around two hundred and fifty billion tons of ice each year, while Antarctica lost about a third as much, in part because there was a small increase in thickness in its eastern section. The great southern blanket may remain intact for another century or so, but after that a collapse of its western section is a real possibility. The ice shelf that floats off that continent has already thinned, in part because the ocean is warmer than it was. When it has gone the waves will nibble at the continental ice itself.

The study of sea levels began in the seventeenth century when Jeremiah Horrocks began to measure the ebb and flow of the River Mersey. A century later his work was extended by William Hutchinson, who also lived on that estuary's banks. He recorded the heights and times of high water for thirty years. His measures were used to generate the first tide tables and were so accurate as to remain useful as a baseline today. After the great East Anglia flood, research into the topic was reborn. The Bidston Observatory on the Wirral Peninsula became the headquarters of a new UK National Tide Gauge Network. That has now taken the ferry across the Mersey to join Liverpool University, the British centre for GLOSS – the Global Sea-Level Observing System.

Changes in level from historical records are hard to interpret for in some places the land itself heaves itself upwards because of a rebound from the lost ice sheets, while in others the opposite is true as the continents tilt on their journeys across the globe. Changes in wind and weather also lead to periods of high or low tides that obscure global patterns, as do patterns of coastal loss or gain. Even a massive hotel built close to an established tidal gauge can depress the surface and give a false appearance of a rising ocean. However, GLOSS now has two thousand stations across the world, many armed with sensors that allow tiny rises or falls in ground level to be allowed for. The information that emerges is complemented by more from satellites that scan the oceans themselves.

Satellite measures taken between 2003 and 2010 suggest that the loss of glaciers alone has led to an increase in sea level of around a millimetre and a half in that brief period. Nobody knows whether the ice sheets will continue to disappear at the present rate and no one can be sure what their contents will do to the shoreline reach but history has some useful lessons. A survey of global sea levels based on the deposition of tiny marine fossils in relation to shifts in temperature as assessed by the composition of stalactites laid down by water slowly oozing in from outside, and by ice cores close to each pole, shows a clear fit with climate. Over the past one hundred and fifty thousand years the level of the sea follows the patterns of polar ice. A decrease in the Greenland sheet is followed by a rise in sea level a century or two later; while Antarctic melting has a larger effect on the tides, but with a delay twice as long.

The seas will rise for other reasons. The Argo system is an array of three thousand floats scattered over the world's oceans. Each sinks to two kilometres down and rises every ten days to broadcast the temperature in the deeps. They have been in place for no more than a decade but already hint at the real increase in the amount of energy stored in the great waters.

Even if emissions are eliminated by 2100 (and few imagine that this will happen) the temperature of the Southern Ocean will rise for a thousand years and the seas are bound to swallow yet more Antarctic ice. Nine-tenths of the energy accumulated by global warming is soaked up in the seas; a figure forty times greater than that for the atmosphere itself. As they warm, the waters expand, and rise.

The Lord explained to Job that the level of the oceans is fixed, for at the Creation he had told the tides that 'hitherto shalt thou come, but no further: and here shall thy proud waves be stayed'. From the time of the Exodus to the twentieth century he kept his promise, for high tide mark was more or less stable, but since the early 1990s there has been an average rise of around three millimetres a year. That has led experts to warn that the oceans will rise by several tens of centimetres by 2100, in what they call 'business as usual', a global economy that produces no more carbon dioxide than it does today. That may be optimistic, for until the recent slump, the substance was being pumped into the air faster than ever before and the new addiction to gas, shale gas most of all, may cause another surge.

Some experts now feel that the earlier estimate of fifty or sixty centimetres by 2100 is now too low, and suggest a figure three times as great. Those of an apocalyptic frame of mind hint at a five metre rise within the next five centuries – a shift as fast as that at the final collapse of the last Ice Age. That would mean catastrophe. For the first time in history, more than half of humankind lives within sixty kilometres of the ocean. Two thirds of all cities of more than two and a half million (places the size of Greater Manchester) are on the coast, as are almost all its megacities, and many are at risk. In India, a metre rise – well within the bounds of possibility – means that forty million people would have to move.

As oceans surge, in some places the land sinks to match. Glacial rebound means that the British Isles are rotating around a north-south axis so that the coast of Scotland rises while eastern England tips into the sea. Elsewhere, the problem is man-made. The great deltas of the world, from the Mekong to the Nile, teem with farmers who drain the land to irrigate their fields or quench their thirst. The coast near Tokyo has subsided by five metres while not far from Bangkok the sea has moved a kilometre inland and has marooned telegraph poles in the shallows. Dams make the situation even worse, for sediment accumulates behind them and no longer settles at the mouth to build up a delta that would keep the rising waters at bay. The Aswan Dam means that for the first time since the days of the pharaohs a whole generation of Egyptians has never seen a flood. Peasants who once grew just a single crop a year now harvest three, helped by thousands of kilometres of canals that steal the river's riches. Nature, as ever, has fought back. The Delta sinks by a centimetre each year and barnacles and other marine creatures have made their way upstream. Soon, the great river will no longer reach the Mediterranean and vast areas will turn to desert.

The Thames Barrier hints at the future closer to home.

It was raised in anger for the first time on the thirtieth anniversary of the 1953 Essex flood, and then three times more in the 1980s. In the next decade it lifted its caissons thirty-five times, and in the first ten years of the new century eighty or so. At the present rate of change the barrier will be overwhelmed by 2060 while, without extensive work on their own defences, Bristol, Maidenhead and Hull will also be at risk. After New York's Hurricane Sandy, there are plans to build a five mile barricade twenty feet high with gates to allow in ships, at a cost of fifteen billion dollars.

The Thames – like all rivers – has flooded many times as the rain pours upon the land. Its overall flow has been more or less stable over the past century and data from York (long a martyr to the River Ouse) hint that even over four centuries British floods have not got much worse on average. Even so there have been real peaks and troughs. Almost all the peaks fit – like those of the Nile – the patterns of weather over the North Atlantic, for that is driven by difference between the extent of high pressure over the Azores, and low over Iceland; a measure that itself responds to the extent of Arctic ice. 2012 was the second wettest year in Britain since records began more than a century ago. We have moved into the part of the cycle that makes for rain, and face another real increase in flood risk as a result.

The need to guard against the deluge to come is as urgent as it was at the time of Noah. Whether his descendants are willing to make the effort to do so is not certain. The present British government has cut down the amount spent on coastal defences and has ignored the advice of experts in the hope that they will get away with it at least for the brief moment that their party stays in power. They may be right. Insurance companies, too, are more and more concerned about the risk. Around two million British homes and commercial properties are already in danger of inundation and the Association of British Insurers estimates that the figure will rise by a third within the next thirty years. The cost of damage – and of cover – has increased by three times in the past decade and is set to soar even higher.

Equitable Life, the company for which Buzz Aldrin, Muhammad Ali and I paraded ourselves, is proof of how narrow is the path between triumph and disaster when making bets against Nature. Its main business was not with natural catastrophes, but in the relative certainties of life, death and retirement. Unfortunately for its shareholders, it misjudged today's huge increase in life expectancy (now a decade longer than that predicted in the 1970s) and offered its investors returns which, in the end, it could not meet. In its attempts to stay afloat, the Equitable – founded in 1762 and with a million and a half policy holders – transformed itself, in effect, into a gigantic Ponzi scheme. At the time of our television campaign in the 1990s it had, unknown to any of us, already misread the balance of risk and opportunity. To disguise the fact that the company was on the ropes, new deposits were used to pay out on older policies. By 2001, Equitable Life had a hidden deficit of around five billion pounds, and soon came the crash. Tens of thousands of investors saw their nest eggs disappear.

As the waters rise, the rains fall and the numbers who face

calamity increase, the profits from flood insurance have collapsed. Faced with the increase in British claims over the past decade, premiums have risen, payment limits are in place and some people are refused cover altogether. In the United States, France and elsewhere the State acts as an insurer of last resort, with a fund to pay those whose lives have been wrecked by natural catastrophes. Britain, alone in the developed world, has no such safety net. In 2013 a temporary pact between the industry and government that insurers will subsidise the costs of the properties at highest risk if the authorities spend money on flood defences will run out. In addition, officialdom has refused loans to the insurers to cover their costs until they have built up their reserves. As the system begins to fail, those unable to pay for cover are in real danger of ruin by the vagaries of nature. Time, perhaps, for every citizen to build their own financial Ark or, if they cannot afford that, to put their trust in God.

CHAPTER SIX

TO THE ENDS OF THE EARTH

William Blake, *God Writing upon the Tablets of the Covenant*

For he said, I have been a stranger in a strange land.

EXODUS 2:22

The second book of the Bible is the tale of a journey. Exodus tells how Moses led the enslaved Children of Israel in a march out of Egypt, across the seas and the deserts, through starvation and revolt, to Canaan's side. There they found the homeland promised long before to Abraham. On the way they received the Tablets of the Covenant and the Ten

Commandments and built a Tabernacle to ensure that God would dwell with his chosen people for ever.

The story, with its ten plagues, the death of the firstborn, the parting of the Red Sea, and the burning bush is among the most familiar in scripture but unlike some of the Good Book's other tales is not matched by any evidence that the events recorded ever took place. If they did, the emigrants would have made an impressive sight, as six hundred thousand Israelite men of fighting age and their families – around two million people – marched, with their livestock, across a hostile landscape. They paused several times on the way (and one camp lasted for forty years) until the next generation was at last allowed to proceed to a land that flowed with milk and honey.

When such a Great Trek might have been, if it happened at all, is just as uncertain. Some rabbinical scholars place it at 1313 years before the birth of Christ. Others argue from historical data that it took place three hundred years earlier. Many of the sites named in the Bible as way-stations were first occupied much later than that and Pharaonic records have no mention of such a dramatic episode, which would have reduced Egypt's population by half.

When the Children of Israel entered the Promised Land they found it populated by alien races. They were enjoined to destroy them. The Lord had said: 'For mine Angel shall go before thee, and bring thee in unto the Amorites, and the Hittites, and the Perizzites, and the Canaanites, the Hivites, and the Jebusites: and I will cut them off ... And ye shall dispossess the inhabitants of the land, and dwell therein: for I have given you the land to possess it.' The migrants, says

Exodus, obeyed those instructions to the letter. In truth, some of the cities said to have been demolished did not then exist and the relics found in Canaan itself show no sign of an explosion in numbers when the alleged multitude arrived. They suggest instead that Judaism grew within that decaying nation's own boundaries.

The history of the Bible's second book is itself disputed. It appears to be an edited version of earlier documents brought together half a millennium before the Common Era by exiles in Babylon who wished to emphasise an identity separate from that of their captors. It was as much a political symbol as an account of real events. The Book was written to preserve the tale of a tribe whom God had identified as a 'stiffnecked people', a race set stubbornly apart. It binds this chosen group – the sons and grandsons of Abraham – to a supreme deity, gives them a unique status in his eyes, and provides them with an eternal home.

Since those days, the history of the Jews has been one of invasion, occupation, expulsion and exile from one country after another. That has, for many, led to a hunger to return to the Promised Land, the biblical kingdom where they have their roots. They are enjoined to do so in the memorable words of the Psalms: 'If I forget thee, O Jerusalem, let my right hand forget her cunning. If I do not remember thee, let my tongue cleave to the roof of my mouth.'

The story of the Exodus is still a potent symbol and the Middle East is now torn by argument about the legitimacy of the state that claims to descend from the exiles. Its emergence involved repeated conflict, and the war goes on. Just before it was established, a band of displaced European Jews travelled to Palestine from France aboard a battered steamship, the *Exodus 1947*. The vessel was blocked by the British, who were desperate to avoid a repeat of the Arab riots that had followed an earlier wave of immigration, but its failure to make the shore was a propaganda coup for the Zionists and in time most of its passengers found their way to the new state by other means.

The claim of a God-given territory has been just as contentious in other places. Herman Melville wrote in *Moby-Dick* that 'We Americans are the peculiar, chosen people – the Israel of our time, we bear the ark of the liberties of the world.' Two centuries before his day, the Puritans saw their own flight from the Old World as an equivalent to the escape from Egypt. Their legal code was based on the Bible: 'The Word of God shall be the only rule to be attended unto in organizing the affairs of government in this plantation.' For them, Thanksgiving was equivalent to the Day of Atonement and the nation which emerged from their efforts still holds reminders of their creed. Yale bears on its insignia an image of the breastplate of the High Priest of the Temple, while the first version of the Great Seal of the United States showed the Jews about to cross the Red Sea (the story that Hebrew might have been adopted as the nation's official language is, alas, apocryphal). In fine biblical tradition, the Puritans – those chosen people – persecuted members of other sects as soon as they had a homeland of their own.

Concrete evidence about the origins of the Jews is fragmentary at best. Their shared pedigree emerged from the ruins of Canaan, Israel's decayed predecessor, which occupied the coastal plain and stretched into the hills to include Jerusalem. By the Late Bronze Age Canaan was in decline. As time moved on and iron replaced bronze, the state began to break apart. Schismatic groups started to demand a separate existence. The bones of pigs are found in the remains of many Canaanite sites, but in time they disappeared from middens in the highlands, perhaps as evidence of the emergence of a new culture. By the late Iron Age those villages had much increased in size. Soon, the Israelites were a political entity and by a thousand years before Christ were in a more or less permanent state of war with their neighbours, and – now and again – within their own ranks.

Sometimes they won, but quite often they did not. In around 1000 BC David is said to have reigned in Jerusalem over a nation that united the northern kingdom of Israel with the southern state of Judea but no generally accepted evidence of his court has been found, although substantial remains show that the city was by then an important centre of administration. The two Books of Kings detail the endless quarrels and reconciliations between the Lord and his chosen people as the latter turn, again and again, to false gods. The divine being punished the apostates without mercy: 'Behold, I am bringing such evil upon Jerusalem and Judah, that whosoever heareth of it, both his ears shall tingle ... and I will wipe Jerusalem as a man wipeth a dish, wiping it, and turning it upside down.'

To show his displeasure, in 722 BC the Lord enabled the Assyrian potentate Sargon to smash up the northern kingdom's own capital and expel its inhabitants. Two hundred years later came the Babylonians under Nebuchadnezzar, who destroyed the Temple in Jerusalem and shattered its adherents' confidence that God had chosen the place as a permanent home. Many Jews were exiled to Babylon itself. On their return after the victory of the Persians in 539 BC they brought with them a purified and strengthened confidence in their own creed. Soon they built the Second Temple and promoted symbols of identity such as circumcision. The Persians were in the fullness of time defeated by Alexander the Great and for a period the Nation of Israel fell under Greek influence. After a variety of internal upheavals and schisms the Romans took power and appointed Herod as the leader of the local province. That led to revolt and to a revenge of the conquerors, who destroyed Jerusalem, most of the Second Temple included. Many of the city's natives fled.

Such upheavals led to the emergence of new Jewish communities to add to those already present around the Mediterranean. Others were scattered as far to the east as India and even China. Legend chased the Jews across the globe and various African tribes are said by some to be their kin, perhaps the descendants of the Lost Tribes. Much of this is fantasy.

Some of the exiles were lost from sight, some kept to their ancient practices in an alien nation, and some managed, much later, to return to what they still viewed as their Promised Land. Their history is full of gaps and has become confused by exaggeration and myth, but now the double helix has followed the family tree of the Children of Abraham to its roots in Africa, to its trunk in the Levant and along its many branches to the ends of the Earth.

Sometimes the molecule confirms the tale set forth in the Bible, but often it does not. Whatever its complexities, the Jewish past as revealed by science has a message for the whole world; that throughout history, humankind has suffered exodus after exodus. DNA also shows that *Homo sapiens* has again and again been on the edge of extinction, with bands of pioneers in flight across a perilous landscape in search of a new home. Our genes tell of exile, loss and disaster. They speak in addition of the universal power of sex even in the face of those who try to keep their chosen peoples apart.

The sole survivors of the biblical deluge were a husband, a wife, their three sons and their spouses. From those few loins sprang, some say, today's multitudes. Should they be right, modern humankind would be reduced indeed. Any population that goes through a bottleneck pays a penalty, for it loses diversity as the genes of those who do not survive disappear into the grave (or the deeps). After the Ark's arrival on Ararat, just ten copies of the double helix – two from Noah, two from his wife, and six from the three spouses of his sons – were left to populate the globe. Such an event would have led to the random loss of certain variants and to great shifts in the incidence of others.

Exile, isolation and loss are still powerful themes. The only eighteenth-century English novel much read today tells the story of another shipwreck. Its sole survivor finds a new insight into the sacred and gains a sense that a sojourn in the wilderness has changed his life.

Daniel Defoe's tale is based on a true story. The island of

Juan Fernández, three hundred miles off the coast of Chile, was renamed fifty years ago by a publicity-conscious government as Robinson Crusoe Island. There, in 1704, the Scot Alexander Selkirk demanded, after a quarrel with his ship's captain, to be marooned upon its deserted shores. He managed to survive on the flesh of feral goats and lasted for four years until he was picked up by a vessel whose commander described him as 'a man Cloth'd in Goat-Skins, who look'd wilder than the first Owners of them'. Once back in England Selkirk gave an account of his adventures to the *Spectator*. It aroused much interest and in time was used by Defoe as the basis of his novel *Robinson Crusoe*. That work has remained a best-seller to the present day.

Defoe's hero bemoans the fact that he had, for most of his stay, no more than a parrot to talk to, while Alexander Selkirk himself recited biblical verses to keep alive his ability to speak. Language connects us to the human race. Crusoe's first task when he came across Man Friday was to teach him English (that done, he converted him to Christianity). The intelligent savage picked up both talents with remarkable ease, but not all of us are so lucky, for some people find it hard to deal with speech. Infants with 'specific language impairment', as the condition is called, find it hard to compose complete sentences or even to pronounce simple words. Some of the many variants of the condition have a genetic basis. In most places, around one person in a hundred has the problem, but a third of the six hundred inhabitants of today's Robinson Crusoe Island are affected, more than anywhere else in the world.

The island stayed empty for a century after the departure of Alexander Selkirk and in time became a penal colony. It was abandoned after a revolt by the prisoners, but was repopulated after 1877 by a small group of stragglers from Chile itself.

Its language problem comes from an accident of history not unlike that faced by the descendants of the Ark. All those who have inherited the condition descend from one of two brothers among the founders, one of whom must have carried the gene responsible. Today's problem is the legacy of an ancient biological mishap; a rare gene which suddenly became common because it was by chance borne by a single member of a small group that later much increased in number.

Many places cut off by oceans, by deserts, by mountains or by social barriers have had such an experience of exile followed by isolation. Often they have a high incidence of otherwise rare genes (some of which cause inherited disease) which have, as on Juan Fernández, become more abundant through simple accident. A huge new survey of variation in the DNA of a thousand people across the globe shows that most populations share the same set of frequent variants, but that the many rare versions are usually confined to one region or ethnic group as a further hint of the random actions of the hand of history in a small and cut-off population. Sometimes, as on Robinson Crusoe Island, the written and the biological records complement each other, but in most places the tale told in the double helix is all that is left.

The Jews themselves show the power of random change. The Ashkenazim are now the largest group and comprise around eight tenths of the world total of fourteen million. The contrarian author Arthur Koestler claimed that they were scions of the Khazar Empire, which for a time dominated Georgia and Southern Russia. In the eighth century, in an attempt to unify its peoples, its aristocracy did indeed adopt Judaism (although that did not last long). Romantic as Koestler's notion might be, Ashkenazim have few connections with the Caucasus. Their immediate ancestors started their European journey in Italy. A few then settled in the Rhine Valley and, around the tenth century, began to move further into what is now Germany, and eastwards into the rest of Central Europe. There the population exploded, from fifty thousand in the fifteenth century to two hundred times as many by the end of the nineteenth.

The Ashkenazim have a history more remarkable than that imagined by the author of *Darkness at Noon*. On the journey up the Rhine, as on that from mainland South America to Robinson Crusoe Island, a moment of near-extinction left a permanent mark. Their mitochondria went through a bottleneck almost as severe as that at the time of Noah. Around half of today's Ashkenazim share descent from just four women, the number on the Ark. The molecular clock of mutations hints that the quartet lived in around the twelfth century. That accident of history still resonates in the health of their descendants. They have rather high frequencies of around twenty otherwise rare inherited illnesses (such as the nerve disorder Tay-Sachs disease, carried in single copy by one in thirty among them, ten times its incidence in the general European population). By simple chance, one or more of those few female founders may have carried hidden copies of those errors that spread to thousands as the population grew. The mutations that lead to inherited breast cancer are also almost identical among Ashkenazim, for just the same reason. A moment of demographic near-disaster dictated the future for millions.

Other exiles from the Promised Land can trace their descent to even smaller numbers of women. Half the Mountain Jews of the Caucasus descend from a single female, as do many among the communities of Mumbai and of Cochin (now known as Kochi) in India.

On the global scale, too, patterns of DNA variation show that the pilgrimage of men and women, whatever their religious affiliation, from their native Africa to the ends of the Earth must have involved whole fleets of allegorical Arks. Some of the vessels made it to safety and those on board flourished in a new homeland. Many more must have foundered before they reached a refuge, with the loss of all hands, and all genes. The most remote places have less diversity than our native continent as a result.

We live in an age of fecundity greater even than that of the Israelites in Egypt. In their day the world held, at a guess, fifty million people. Then numbers began their inexorable rise. They took from the Big Bang (or, for literalists, since the day of the Flood) until 1927 to reach two billion, until 1974 to multiply that figure by two, and until 1999 to add yet another two billion. Nobody before the twentieth century had seen the human population double, but today's octogenarians have seen a threefold increase. The year 2011 saw the seven billionth baby. Today's growth rate is half that of the 1960s, but a new France still arrives each year. Numbers will peak at around ten billion in 2100.

Huge as that figure is, patterns of inherited variation show that the whole human population is still in recovery from a series of ancient disasters and that, for much of the past, *Homo sapiens* was an endangered species.

Chimpanzees hint at what a hard time our ancestors must have had when they set off on their global journey. The animals' present plight is dire. Just a century ago two million chimps lived in Africa but now they are reduced to less than a tenth of that number and may be gone from the wild within a century. Even so, in terms of diversity, the two primates are mirror images. Since the split from their common ancestor with ourselves some eight million years ago, chimpanzees have retained far more variation than have humans. That hints at a human bottleneck. We also possess vast numbers of individually rare genetic variants as further evidence of a forgotten Crusoe experience: a population explosion from a tiny base as recently as five thousand years ago, not long before the earliest events recounted in the Bible (and close to Archbishop Ussher's estimate of the date of the Creation). The genes show that chimpanzees have always been rather common while men and women, in spite of today's profusion, have flirted with extinction again and again. Whatever our present billions, averaged over the past half million years or so, our own population has in effect been no more than ten thousand, that of a small town today.

As humans filled the world the general picture was of

moderate abundance – perhaps hundreds of thousands – in some places, but punctuated by severe reductions as bands of pioneers moved to new and empty lands.Our expansionist urges have made us into a diminished primate.

Humans of modern form left Africa rather more than a hundred thousand years ago, several tens of thousands of years after they first appeared there. Because of the cold they did not make much progress until around sixty thousand years later, by which time they had reached western Asia. Australia was settled ten thousand years after that and the first known Britons of modern form (who chose to live in Torquay) did not arrive for another ten millennia. New Zealand remained uninhabited until around AD 1250, and some of the remote Pacific islands were empty until just a few centuries before the present.

The biblical tale of an ancient group of pioneers has an echo in reality. Comparison of the DNA of modern populations on the flight from their roots in Africa, from there to Europe and to Asia, and onwards to remote islands such as Tahiti or New Zealand, shows that, on every step of the journey, more and more diversity was lost as proof that very few people made their way to each of those new-found lands.

Twenty-five years ago, with the help of a young Iranian physicist, I made the earliest (and entirely forgotten) attempt to work out the size of the first bottleneck on the global journey; that out of our native continent. We had information on just a short length of DNA sampled both within Africa and outside it and used the figures to calculate how small the emigrant population must have been to enable the accidents of sampling to cause the observed drop in diversity. We came up with the remarkable result that just one couple might have made the first exodus. The total would be larger if the emigrant group stayed small for many generations; for example, six individuals for two hundred years would have the same effect. By coincidence, that is just the same number as there were passengers on the Ark. Now, with information on a quarter of a million sites across the DNA in far more people and with a much more refined analysis, the figure for a single generation has, in an another reminder of the Genesis story, risen to six.

The journey from our homeland got off to a difficult start. The experience was repeated again and again as small bands struggled onwards to the coasts of the Atlantic, the Indian Ocean and the eastern shores of Asia.

For all that time a land without people was waiting. Twenty thousand years ago, men and women began at last to stray into Melville's Ark – the Americas – from Siberia. In time they filled its twin continents, but for thousands of years its peoples were cut off from the world. Then, everything changed. On October 12th 1492 Spanish ships anchored off an island in the Bahamas. Those vessels, and the many that followed, brought waves of European and African genes to the New World. In an evolutionary instant, what had been a remote and reduced outpost of humankind became a microcosm of what the future holds for us all.

The Jesuit José de Acosta spent most of his life in Latin America and crossed the continent again and again. In his 1590 book *The Natural and Moral History of the Indies* he remarked upon earthquakes and the tsunamis that followed them, studied altitude sickness in the Andes and described how the locals used cocaine. On his travels, which took him to Peru, Bolivia, Chile and Mexico, the priest noted the physical similarity of the natives to the peoples of the Far East. De Acosta came up with a radical suggestion: that the first Americans had migrated from Asia across a land bridge, now submerged.

He was right. Apart from the few stragglers who made it, much later, to distant places such as Tahiti, Hawaii or New Zealand, the entry into the New World was the final major step on mankind's exodus from Africa.

The Bering Strait – once the Bering Land Bridge – was not discovered until more than a century after de Acosta's death. The place is named after the Danish navigator Vitus Bering, who had been hired by the Russians to survey the Kamchatka Peninsula at the eastern extremity of Siberia. He sailed through the Strait in 1728 and died there on a second expedition a few years later. Its waters are in places no more than thirty metres deep. In a parallel to the parting of the Red Sea, as the ice sheets grew they fell away altogether, to offer a passage to a band of emigrants.

At the time of the last great freeze, with sea levels a hundred metres lower than now, Beringia was a vast plain a thousand kilometres wide. For much of its history it was a cold, dry and hostile steppe. Now and again sparse forests of birch and poplar sprang up, but the Bridge was a hungry place. Mammoths and sabre-toothed cats roamed the landscape as did a few scattered bands of hunters, who ate what they could kill and moved back and forth between what is now Alaska and Siberia. Around seventeen thousand years ago, a few ventured further into the Americas.

As the world warmed, the waters returned, to give the first inhabitants of the Americas a reduced version of their own Great Flood. The isthmus began to shrink until, around fourteen thousand years before the present America broke its ties with Asia, the people on the eastern shore found themselves on a new continent, marooned between a stormy ocean and a great range of mountains.

Soon, the ice retreated further. The glaciers of the Canadian Rockies shrank and opened a corridor from Alaska into the Great Plains. Another escape route was provided by the narrow plain that still lay along the coasts of Alaska and of British Columbia. Each was a door to a treasure-house beyond. The immigrants trudged onwards, and reached the continent's southern tip within three thousand years.

Almost no relics of that earliest occupation have been found, but a fourteen-thousand-year-old site in Central Texas called Buttermilk Creek has yielded a mass of simple tools, not much different from those then made in the Old World. Buttermilk society was supplanted by the first genuine American culture, the Clovis people (who are named after the town in New Mexico where their vestiges were first found). Their remnants date from around eleven thousand years ago and are scattered across the Great Plains and the southern United States, and their way of life has been identified in Mexico and even further south. Most Clovis sites are in the east rather than the west, evidence that a wave of travellers moved towards the Atlantic coast before they began the trek towards Cape Horn. Their economy lasted for thousands of years. Shaped stones were used as blades and as arrowheads. Using spears launched with throwing-sticks they killed off the mammoths, bison and other large mammals that once roamed the continent.

Once arrived in the grim landscape of the far south the immigrants must have lived in much the same way as did their descendants, the Fuegians as Charles Darwin called them on his visit in 1832. The Yaghan people (to give them their correct title) wore almost no clothes, and survived the bitter cold in small groups huddled around fires (a habit which gave Tierra del Fuego its name). Darwin thought them to be 'miserable, degraded savages' and wrote that 'I could not have believed how wide was the difference between savage and civilised man: it is greater than between a wild and domesticated animal, in as much as in man there is a greater power of improvement.' Be that as it may, the southern tip of Patagonia has a higher density of archaeological sites than anywhere else on Earth, as evidence of a sophisticated and successful ancient society (and Darwin later moderated his contempt).

The New World was a place of small, mobile and isolated bands. The shortage of relics hints at how sparse its peoples must have been. Biology, too, provides evidence of how few made it to their new-found land and of how even fewer struggled towards the final steps of the journey.

American skeletons are less variable than are those of their African or Asian ancestors. Two Native Americans from the same place hence look more like each other than do two people from other parts of the world. In an unexpected echo of the past, the bacteria in their guts are also less diverse than those found in Old World intestines, as a further hint that the internal fellow-travellers suffered a collapse as they passed through bottleneck after bottleneck. Their languages, too, speak of difficult times. Each of the globe's idioms consists of a series of distinct sounds. Their numbers vary from place to place. Africa is the world capital, with a remarkable range of clicks, tonal changes, whistles, guttural noises and so on. English is reduced in comparison, while the native tongues of the New World are even more diminished. At the southern tip of the continent, where Darwin remarked (unfairly) that 'The language of these people, according to our notions, scarcely deserves to be called articulate', speech is at its most truncated.

The best evidence of how hard the journey to – and through – the new continent must have been comes from the double helix. There is a hole in the data, for many Native Americans in the United States have refused to cooperate as they prefer to hold to their traditions of a miraculous emergence on their own territories. Even so, the genes tell the remarkable tale of the first entry into what, long afterwards, became a Promised Land for much of the rest of the world.

To reach its empty landscapes the escapees from Africa were forced, like their biblical descendants, to cross deserts and to climb mountains and to make great detours, not around the mountains of Sinai and the Red Sea, but around impassable obstacles such as the Himalayas and the Indian Ocean. A comparison of today's genetic variability in populations from across the globe with the walking distance from

Addis Ababa reveals a good fit between the two, with the peoples of the most distant places much less diverse than those who stayed close to home. The natives of South America are the most reduced of all, with a quarter fewer single-letter differences in the DNA than found among sub-Saharan Africans. Their impoverished state shows that the continent's founders were few in number even when they arrived in Alaska, and lost yet more diversity on the rocky road to the south.

Beringia was America's Ark. The genetic shift in the Americas compared to the Old World suggests that the native population of the whole continent, North and South, descends from fewer than a hundred people. The double helix binds most of today's Native Americans not to Siberia but to the peoples of Kyrgyzstan and the Buryat Republic, thousands of kilometres away. Those remote nations share more ancestry with Americans than do today's Siberians, perhaps because of movement within Asia since the days when the pioneers struck to the east.

After the first wave of colonisers, immigrants continued to trickle into the New World. The natives of the Aleutian Islands speak a language distinct from other Native Americans and have genetic ties with Siberians and also with Greenlanders, who may have spread there from Alaska across the far north (an idea supported by the five-thousand-year-old frozen corpse of a Greenland Inuit, which shows closer affinity to Siberians than to other inhabitants of the Americas). Canada's Chipewyan Indians are yet another group for they have ties with the ancestors of today's Han Chinese.

Y chromosomes lose variation on the road to the south faster than do other segments of the double helix. The leaders of the Aztecs and the Mayas were as avid for intercourse as were their Old Testament equivalents. A primary wife provided them with official heirs, but they were allowed many secondary partners. Some fathered enormous families. As a result other males – because they were sacrificial victims, were killed in wars, or were too poor to attract a mate – had no children at all. Their DNA was at a dead end, which meant that the bottlenecks bore more upon men than upon the opposite sex.

For Native Americans, male and female, the arrival of the *Niña,* the *Pinta* and the *Santa María* marked the end of ten thousand years of solitude. The once empty continent changed for ever. Men and women poured in from across the globe. They brought not a trickle but a torrent of DNA. The post-Columbus clash of ancient with modern put paid to mankind's long era of random change. The peoples of the Americas began to merge with the invaders until their adopted landscape became the most diverse continent the world has ever seen. Now, the whole planet is on the edge of an age when admixture is all.

The Spaniards, like those who had made the Exodus from Egypt, believed that they were on a mission to a land destined by God to be their own. Columbus himself had quoted an apocryphal book of the Bible in his attempt to persuade the nation's rulers to finance an expedition to India with a journey to the west. The voyage was an attempt to fulfil Isaiah's prophecy that, to complete the diaspora and bring about the Second Coming: 'Surely the isles shall wait for me, and the ships of Tarshish first, to bring thy sons from far, their silver and their gold with them.' Tarshish was, in this view, a mystic city in Ophir, in the Far East, and Columbus was influenced by the hope of an era of bliss – the recapture of Jerusalem for Christianity included – that would emerge when the Word of God had at last reached that remote spot, and had filled the world. The riches that awaited him ('Then shalt thou lay up gold as dust, and the gold of Ophir as the stones of the brooks') would pay for the rebuilding of the Second Temple.

On arrival in what they still saw as an outpost of Asia, the Spanish conquerors denied even that the local Indians belonged to the same species as themselves. They could, like the Perizzites, Canaanites, Hivites and Jebusites of old, be killed with a clear conscience, to allow God's will to triumph.

José de Acosta disagreed. He championed the notion that Amerindians should be regarded as human even if he did feel that they were ruled by Satan in the form of their own murderous gods. After much persuasion the Conquistadores admitted as much but that did not stop the slaughter. Columbus said of the Taino people of Hispaniola that they were 'innumerable, for I believe there to be millions upon millions of them'. The true figure was several hundred thousand, but within half a century just a few hundred were left, the rest worked to death in the new gold mines, or killed off by infectious disease. The first great epidemic – perhaps influenza – struck just a year after the Europeans arrived. Pestilence travelled faster than did soldiers, who, as they moved further into the continent, encountered peoples already almost destroyed. The population of Central America fell from twenty-five million in 1518, the year before Cortés began his conquest of the Aztecs, to seven hundred thousand a century later.

Some silent witnesses hint at how many must once have lived in the transatlantic Canaan. Large parts of South America now impenetrable were not temples to untamed wilderness, but open landscapes that buzzed with activity. The Atlantic forest of Brazil, at the time of Darwin's visit a dense jungle, had once been burned every few years by its inhabitants. The evidence lies in the carbon deposited on the floors of nearby lakes. Much of the Amazon rainforest is also new. Before Cortés, the landscape was scattered with villages, the country farmed with care. Banks and ditches hundreds of metres across and approached by long avenues remain. Together they supported millions. After the collapse of the human population they disappeared under rampant vegetation.

As the devastation went on, more and more migrants were pulled – or pushed – into the New World. Bartolomé de las Casas in his *A Brief Account of the Destruction of the Indies* of 1542 claimed that the Indians were unsuited to hard work and that they should be freed from slavery. They were – but they were replaced by Africans. The earliest arrivals from that continent landed on Hispaniola in 1501, less than a decade after the first Europeans. In the next three and a half centuries, twelve million joined them (which meant that four times as many people with black skins immigrated in that period as did those with white). Only a minority were taken to the United

States while almost all the others went to the Caribbean and to South America.

Sex started at once, and involved all parties. Many of the Europeans who colonised Hispaniola married local women. Cortés himself impregnated a princess and had several other half-native children, one of whom even found a place in Spanish society. Pizarro, too, had children by members of the Inca nobility and, no doubt, many soldiers entered into their own engagements. From the first days of the European arrival the imperatives of biology spilled over the social barriers. That history has a reflection in modern Spain: if one of the Spaniards whose DNA was read off in the global survey of DNA variation across a thousand genomes has a variant shared by just one of the others, there is a 50 per cent chance of that second individual living in South America, as proof that genes crossed the Atlantic from west to east as well as in the opposition direction.

In the New World itself, there was (and still is) a certain status associated with an individual's racial background. Parts of that continent had for many years a taxonomy that classified people by their genetic history, as 'wolves', 'Moors', 'those suspended in air', and other fine distinctions, with prestige dependent on how many European ancestors any individual could claim. Modern Colombia is a microcosm of that process. Nine out of ten of its inhabitants identify themselves as of mixed ancestry, while most of the rest think of themselves as displaced Africans. No more than a tiny proportion claim to be Native American.

The nation's genes reveal a more subtle blending of DNA.

As was the case for the Cape Coloureds, admixture by men and by women was less than even. 90 per cent of the mitochondrial lineages of Colombian cities are of Amerindian ancestry while fewer than one in twenty comes from Europe. Almost half their Y chromosomes, in contrast, have a European origin, as a reminder of the sexual habits of powerful men. Everywhere, the real proportion of European genes in its inhabitants is much higher than that obtained by asking people what group they feel they belong to, for in Colombia, as in other places, mixed-race people with rather dark skins tend to over-estimate their African ancestry (Condoleezza Rice professed herself astonished when she learned that more than half her heritage comes from Europe). She and many others place more weight than it merits on the small set of genes that determine physical appearance.

In the five short centuries since the Europeans arrived, the New World has, thanks to the power of lust, moved from its status as the most diminished branch of the human kindred to its most diverse. The desire for sex has overcome the barriers of colour, culture and creed to brew up a rich biological soup.

That history of gene exchange among groups once defined as utterly distinct has a resonance in the biblical Exodus. Those ancient travellers to a Promised Land were – like the Conquistadores – admonished to keep themselves pure in spirit and in behaviour, and to avoid liaisons with members of lesser tribes ('. . . the LORD, whose name is Jealous, is a jealous God: Lest thou make a covenant with the inhabitants of the land . . . And thou take of their daughters unto thy sons, and their daughters go a whoring after their gods').

The Children of Abraham, like those of Columbus, did not live up to expectations. They began to stray almost at once. Ezra (who lived in the fifth century BC) was outraged: 'The people of Israel, and the priests, and the Levites, have not separated themselves from the people of the lands, doing according to their abominations ... so that the holy seed have mingled themselves with the people of those lands ... And when I heard this thing, I rent my garment and my mantle, and plucked off the hair of my head and of my beard, and sat down astonied.'

The Prophet's fury was justified. The double helix shows that the history of his people has, from its earliest days, been tangled indeed, with evidence of exile, return, shifts of identity, and copious admixture with those who surround them. Just as in Melville's 'Ark of the liberties of the world', the inhabitants of the many new Jerusalems that sprang up across the globe indulged in plenty of sex with the neighbours.

The habit began in the earliest days. In Judaism membership is passed down the female line and although conversion is possible the process is arduous. Most modern Jews have a Jewish mother but the mitochondria tell a more ambiguous tale about how long, and with what rigidity, that rule has been upheld. The Beta Israel – the Black Jews of Ethiopia – are African in appearance. One legend suggests that their conversion may have come five centuries before Ezra, with Menelik, the son of Solomon (and perhaps of the Queen of Sheba), the founder of the country's main dynasty. His royal lineage lasted, after a convenient shift of allegiance to

Christianity, until the fall of Haile Selassie in 1974. Some claim instead that the Beta Israel emerged from a migrant band of wanderers, perhaps the Lost Tribe of Dan, or even from Christians who objected to the doctrines of the Ethiopian Church and converted to Judaism within more recent times.

Genes show that the Jewish invaders exchanged more than ideas with the Ethiopians. Beta Israel mitochondria resemble those of other Ethiopians, while their Y chromosomes have a tie to modern Jews. The Beta Israel descend not from immigrant women, nor from recent mass conversion, but from Abyssinian females who entered into relationships with the intruders. In spite of initial disagreement as to whether these Black Jews had the right of return, in 1991 there was a huge rescue mission to fly them to Israel.

Such exchanges with non-Jewish populations have gone on within Israel, and across the world, throughout the centuries. Many of their lineages resemble those of non-Jews as evidence of a long history of admixture.

Ezra was particularly outraged that the hereditary priests of the temple should have liaisons with members of other faiths. The genes show that he was right to be annoyed. Around half the cohanim, whose surname identifies them as descendants of the temple priesthood, do share similar Y chromosomes as proof of common ancestry but the rest have an assortment of such things from other males who have sneaked onto the pedigree.

On the larger scale, the picture is just as confused. The DNA of Jews from Western Europe (Ashkenazim included), and from Iran, Iraq, Syria, Italy, Turkey and Greece, reveals some shared links with the Middle East but also has strong ties with that of the non-Jewish peoples amongst whom each has dwelt. European and Syrian Jews taken together show a modest divide from those from the modern Middle East. The Jews of the northern and southern shores of the Mediterranean both trace their origin in part from their fellows who left Israel in Greek and Roman times, but many on the European side have quite close kinship with Italians. That may be in part due to intermarriage, but conversion is also involved for at the time of the Classical occupations of the eastern Mediterranean thousands of pagans abandoned their earlier practices to take up Judaism. Iranian and Iraqi Jews, who claim descent from the Babylonian exiles, have rather fewer alien genes than others, as proof that their diaspora was truer to its roots than most, but the general picture is one of assimilation.

As their numbers grew, the Children of Abraham moved on. The city of Tarshish was probably – in spite of Columbus' belief that it was in the Indies – in southern Spain (Jonah was on his way there when he was swallowed by the whale). There, long ago, settled another branch of the Abrahamic household. The Sephardim, as they were known, began to arrive at the time of the fall of the First Temple from the biblical Sefarad (which may have been Sardis, capital of Lydia, now part of Turkey). Many more moved to Iberia after the Roman conquest of the Levant.

As the Visigoths, the Germanic people who ruled the peninsula in ancient times, took up Christianity, discrimination against the Sephardim increased until, in the end, non-Jews were forbidden even to speak to them. Many left for North Africa. Then, in AD 711, came the Muslim invasion of Spain. Some of the Jewish exiles in the Maghreb welcomed the chance to return to Iberia and as Islamic rule moved onwards Jews took over the government of certain towns (a fact later used as an excuse for persecution). Under the new regime they flourished as bankers, physicians, merchants and the like. Their numbers increased to almost half a million, with plenty of marriages outside their own community, to such an extent that even Ferdinand II, who authorised Columbus' voyage of discovery, had Jewish ancestry (although it was not prudent to point this out).

Soon the situation changed, once again, for the worse. Fundamentalism grew among the Islamic invaders, who began to bully all non-believers. Some among the Sephardim moved to a more tolerant Portugal, or left Iberia altogether. Then came a further triumph of bigotry and, for the remaining Jews, disaster. The Christians fought back against the North African invaders, and won. On the final defeat of the Muslims the fate of the Sephardim was sealed. In spite of his own family history, Ferdinand II, supported by the Church, issued a decree that demanded their immediate conversion or expulsion. Some Jews were burned at the stake, many more left, while others converted, or at least appeared to do so. Thousands fled, some back to Africa and more to the Ottoman Empire, where they were welcomed as merchants and as people of education. They were the kernel of much of the modern Jewish population of Greece and Turkey, which still has biological links with the opposite end of Europe.

In Spain and Portugal, most of those who remained claimed to accept Catholicism. Often that was no more than a gesture that allowed the *conversos*, as they were called, to keep their place in society while in private they continued to follow their own credo. The secret Jews were hunted down by the Inquisition who identified them through careful observation of their habits, such as a tendency to wear clean clothes on the eve of the Sabbath. In spite of such persecution they maintained a hidden identity for many years. In a certain Portuguese village one community of secret Jews married among themselves for five centuries, into modern times. As a result, all but thirty or so of its four hundred present inhabitants trace descent from just one woman who lived during the reign of Ferdinand and Isabella.

Many *conversos* adopted particular surnames that allowed those in the know to continue to discriminate against them. So convinced was Rome of the dangers of non-Christian blood that it passed a law of *limpieza de sangre* – of purity of blood – that lasted until the nineteenth century in the army, and until the 1960s in Majorca, where no priest with a Sephardic ancestor was allowed to celebrate Mass in the cathedral.

The Sephardic contribution to modern Spain is in truth far greater than most of the country's inhabitants, and most Jews, imagine. It shows the extent to which, however strict the authorities, admixture is impossible to control. A fifth of the Y chromosomes of today's Spain are of *converso* origin and in southern Portugal the proportion rises to as much as one in three. In the land of Tarshish, the diaspora lives on in body even if much of its spirit has been forgotten.

On the 3rd of August 1492, just a day after the deadline for Jews to convert, to leave, or to die, Columbus' ships set forth westwards. After a stay in the Canary Islands to pick up supplies, they set off once more. In the Bahamas the explorers met their first Native Americans (who could, Columbus thought, be persuaded without much difficulty to take up Christianity). From there his vessels moved on to Cuba and to Hispaniola. A year later he was back in Spain preparing for a larger expedition, supplied with priests, colonists and soldiers. The Old World was about to join the New.

Ferdinand and Isabella were as a result – and quite inadvertently – the agents of the last great step in the dispersal of the Children of Abraham, for they forced the exiles into a new continent, where they were to flourish. The laws passed by the Spanish and Portuguese for their new Christian Empire across the Atlantic banned *conversos* from the territory. That policy failed from its first day. Several of the crew of the *Niña*, the *Pinta*, and the *Santa María* are said to have been converted Jews, anxious to escape the dangers of Spain. Luis de Torres, for example, was born a Jew and had served as a government interpreter. Columbus took him as a translator, perhaps because he thought that his expedition might encounter some of the Lost Tribes of Israel and that a Hebrew speaker would be useful (de Torres is also said to have been the first European to smoke tobacco). Not long afterwards, several of Cortés' own soldiers were executed on the grounds that they too were *conversos*. Some enthusiasts even claim that Christopher Columbus himself had Sephardic roots although a DNA test of his descendants has failed to confirm the idea.

In time, many more European Jews fled across the Atlantic. By the mid-seventeenth century there were large communities in Brazil (which had the first synagogue in the Americas) and elsewhere. Two centuries passed before many made it to the United States. Recorded Jewish history there begins in 1654 in New Amsterdam, now New York, with a group of refugees who had fled from Recife in Brazil, from whence the Dutch had just been ejected by the Portuguese. At the time of the American Revolution, there were around two thousand within its borders, most of them Sephardim, and many took part in the struggle against the British. The first synagogue opened in Rhode Island in 1790 and was welcomed by George Washington, who referred to the prophecy of Micah that 'they shall sit every man under his vine and under his fig tree'; in other words that his new nation would, he hoped, become a land of tolerance. By the Civil War the number of Jews had gone up by fifteen times. At the end of the First World War there were more than two million in the United States, most of them Yiddish-speakers from Russia and Eastern Europe. They were joined over subsequent decades by many more desperate to escape from the Nazis and, in recent times, by migrants from the former Soviet Union. In their final exodus the Children of Abraham travelled, like those in the biblical account, in millions.

They brought their credo, their culture and their genes. In time, as in Ezra's day, they began to merge into the nation as a whole. By the 1950s in the United States more than half of all Jewish marriages were with members of other faiths or of none. People who identify themselves with that community now represent a smaller proportion of the American population than they did even a century ago and their relative numbers are set to fall further as other groups — Hispanics most of all — expand.

The story of America's Jews is the story of us all; a tale of exile, danger, discord and separation followed, in the end, by the collapse of social and religious barriers. Many have lost their identity, and some have forgotten their roots, but the double helix reminds them of the past. Americans with four Jewish grandparents can be separated with certainty from others with a scan of a few sections of the double helix and even those with just one grandfather or grandmother from that community stand out as distinct. Within the white population of the United States there are three ancestral clusters, from north-west Europe, from the south-east of that continent, and from the Ashkenazim. A set of three hundred variant sites is enough to ascribe every white American to membership of one or more of those lineages.

Perhaps the Patriarchs (not to speak of the security guards at Tel Aviv Airport) would be comforted by the fact that DNA has revealed a forgotten Jewish ancestry for millions. It shows that the people of Israel — like everyone else — have been through adventures and disasters in the search for a new homeland that make those of the Exodus look tame. Ezra tore his hair at the antics of his fellows; but for them, as for all of us, the imperatives of biology far outweigh the demands of doctrine.

CHAPTER SEVEN

THE LEPER'S BELL

William Blake, *Pestilence*

To teach when it is unclean, and when it is clean: this is the law of leprosy.

LEVITICUS 14:57

After his betrayal of Jesus, in a fit of remorse Judas Iscariot returned the thirty pieces of silver to the priests and hanged himself. The priests, dubious about his tainted gift, used the cash to buy a potter's field – a worked-out clay deposit – and offered it as a burial place for strangers and non-Jews, a role it fulfilled until the nineteenth century. The red clay gave it the name of Haceldama, the Field of Blood (although the Acts of the Apostles suggests instead that the name came from the unfortunate fate of Judas: 'falling headlong, he burst asunder in the midst, and all his bowels gushed out').

Many of Haceldama's tombs have been looted but some still contain human remains. One such is the burial place of a family, for the DNA from a first-century sepulchre shows that several of the occupants were related. One of the niches is walled off with plaster. It contains a shrouded male corpse, perhaps that of a priest. Molecular tests reveal the presence of the agent of leprosy. In biblical times that disease was feared above all others and was, no doubt, why the tomb was sealed.

Leviticus, the Book of the Levites, was composed over several centuries from around 1400 BC. It sets out a number of rituals designed to cope with disease. They are, in many ways, a quest for order: a realisation that, whatever its cause, illness was an insult to the system by which the Lord meant Man to live. Many of its verses are devoted to leprosy and to the rules to be followed should it be found: the presence of a 'rising, a scab, or bright spot' on the skin calls for quarantine and if 'the scab spread much abroad in the skin' the priest should pronounce the sufferer unclean. The unfortunate patient then faces exile: 'His clothes shall be rent, and his head bare, and he shall put a covering upon his upper lip, and shall cry, Unclean, unclean. All the days wherein the plague shall be in him he shall be defiled; he is unclean: he shall dwell alone; without the camp shall his habitation be.'

Leviticus, in that passage and elsewhere, is obsessed with

hygiene. Animals were classified as clean or otherwise and any object touched by the latter group (which stretched from camels to ravens) became polluted. Sex, too, was tainted unless strict rules were followed; for a man, even to sit in a chair once occupied by a menstruating woman was forbidden and after an unexpected ejaculation he was impure for the rest of the day. A ritual wash (and now and again a ceremonial shave) was needed after handling a corpse and in other circumstances. Other passages give more practical sanitary advice: 'And thou shalt have a paddle upon thy weapon; and it shall be, when thou wilt ease thyself abroad, thou shalt dig therewith, and shalt turn back and cover that which cometh from thee.' Purity remained at the centre of biblical doctrine. For the Israelites, corporeal cleanliness was paramount, but in the New Testament purity of heart became the key.

Keen on hygiene as the ancient world may have been, it was in truth filled with infection. Disease is a repeated theme in the Bible (although treatment is not: almost the only cures on offer are supernatural). Some of the afflictions were unpleasant. Job complained that he was eaten alive by worms while Zechariah predicted for the enemies of Jerusalem that 'Their flesh shall consume away while they stand upon their feet, and their eyes shall consume away in their holes, and their tongue shall consume away in their mouth.' Most such conditions cannot now be identified although there are hints about some. Tobit, in the Apocrypha, sleeps outside by a wall in which there were sparrows, who 'muted warm dung into mine eyes, and a whiteness came in mine eyes' (he almost certainly had cataracts). In the same way, Jehoram son of

Jehoshaphat may have had dysentery: 'In process of time, after the end of two years, his bowels fell out by reason of his sickness: so he died of sore diseases.'

Jehoram had annoyed the deity by murdering his brothers and sisters and his plight, like that of many others, was ascribed to the anger of the Supreme Being. A whole community might be smitten: 'Because of the wrath of the Lord it shall not be inhabited, but it shall be wholly desolate: every one that goeth by Babylon shall be astonished, and hiss at all her plagues.' That verse was written at the time of exile of the Jews to Babylon in the sixth century before Christ, the period when the city was the largest in the world. It sucked in migrants from afar to become the first settlement to number two hundred thousand people. That link of mass movement and the emergence of a vast metropolis with infectious disease has a modern air, as the spread of the human immunodeficiency virus from the West African bush to the Congo's new cities and onwards across the globe reminds us. The first days of the Israelites marked the start of the age of epidemics.

The leprosy of Leviticus may not have been the malady we now know by that name. Because of the opprobrium attached to the term, the condition is now called Hansen's disease after the discoverer of the bacterium responsible. The biblical version was probably a complex of skin infections such as ringworm, psoriasis and boils. Certain lepers were said to be 'white as snow', which might mean that albinism (not, of course, contagious) was also involved. The matter is confused by the statement that leather, clothes and even houses might suffer from leprosy and by detailed instructions as to how they must be purified and, if that failed, should be destroyed or demolished. The term may sometimes have referred not to physical illness but to a general sense of rejection and divine displeasure, with no biological implication at all.

The notion of health as a gift and of sickness as chastisement lasted from biblical times almost to the present. Disease was a just punishment for sinners. Great plagues appeared from nowhere to strike the impure and when the correct rituals had been completed ebbed away. Most medical treatments were useless or worse and not until modern times did doctors begin to cure more people than they killed.

Medicine is now more effective than when its sole remedy was that 'the leper ... shall cry, Unclean, unclean'. Its progress has been spectacular. Even in the first half of the twentieth century, infectious disease was the developed world's main cause of death. The influenza epidemic of 1918 in a single year put an end to twice as many people as AIDS has killed over the past four decades. Many of those plagues have gone (although they may, needless to say, come back).

Sewers and social progress have played a larger part in that success than has science, but scientists have at least begun to understand the origin and nature of contagious diseases. Where do they come from, why are some severe and some mild, and why is each so distinct in its symptoms? Why are certain conditions infectious and others hardly so, and why are some lethal in hours while others take years to kill? Why should sudden outbreaks happen after decades of calm and then disappear with equal speed? The biblical view of illness as conflict with the order set by God hints at a deeper truth;

that all diseases are a struggle between two (or more) parties. They emerged not as supernatural punishments but from changes in our way of life, many of which took place in Old Testament times. The move from the hunting ground to the fields, from village to city, and from a lifetime in the same household to an era of mass movement each led to great shifts in the patterns of infection.

Leprosy was among the first conditions to leap onto the bacterial bandwagon. An illness that sounds much like it is recorded in the Sushruta Samhita, a text of Ayurvedic Medicine written in India around 600 BC, while skeletons show that the same disease affected humans four thousand years earlier than that. The Indian work recommends treatment with the oil of a certain tree (it also hints that suicide, normally forbidden to Hindus, was not a sin where lepers were concerned). The oil is still used by herbal healers but does not help.

The illness has an intimate relationship with man. It infects no other animal except, bizarrely, the American nine-banded armadillo (which picked it up from European immigrants and within which the bug can thrive because of its low body temperature). Until not long ago the bacillus could be cultivated only in human cells but now mice with weak immune systems and armadillos themselves are used in research.

Leprosy's main target is the nervous system rather than (as the Levites thought) the skin. Its agent, a bacterium called *Mycobacterium leprae*, attacks nerves in the face, fingers, testes and other cool parts of the body. The immune system responds with a failed attempt to wall off the invaders. That leads to inflammation, to the accumulation of clumps of white blood cells and decayed tissue, and to the destruction of nerves, skin, the lung surface and more (the popular idea that fingers and toes drop off is incorrect, but they may become shorter as bones are destroyed). As time goes on, the muscles become weak, and patients lose the powers of taste and smell. A feeble voice and a 'lion face' are a further scourge. A few men even lose their testicles and grow breasts. Putrid flesh may generate a foul odour that defines the sufferer as 'unclean'. Until not long ago many among the afflicted were so shamed by that label that they denied their illness and avoided treatment.

The agent of leprosy is protean in its effects and patient in its work. The average incubation time, from infection to diagnosis, is five years but in some people may be eight times as long. Its cells divide no more than every two weeks or so, and the number of bacteria present when the condition is finally diagnosed is greater than that of almost any other illness. Nine out of ten people are resistant to attack, and many of those infected are quite unaware of the fact. In spite of the Levites' ancient terrors the condition is not very contagious. Its agent is in the main passed on by bacteria sneezed out by its bearers, while transfer by touch or from mother to foetus or to infants from breast milk plays a lesser part. There have even been cases in which, in a revenge of the armadillos, people have picked up the disease from those animals. Around eighty Americans a year are infected in this way, most of them in Louisiana and Texas, where the locals hunt, skin and eat such creatures.

Infectious or otherwise, the power of biblical injunction meant that for most of history lepers were forced 'without the camp'. They were impressed into special colonies, or were obliged to carry horns or rattles to warn of their approach. The colonies or 'lazarets' (named after the infected beggar Lazarus, gathered into Abraham's bosom) were on islands or in isolated valleys, where, quite often, those forced to live there were abandoned. In Hawaii, whence the disease was brought by Chinese immigrants, the Segregation Law of 1865 banished patients to a remote peninsula on the island of Molokai. There, chaos and crime ruled until an itinerant Belgian missionary improved matters. The place remained as a quarantine site until 1969. Japan kept its policy of isolation into the 1990s, with hundreds shut away against their will. Such was the terror of contagion that some countries printed special 'leper money' for fear that to allow patients to handle cash in general circulation would cause the disease to spread. A few lazarets remain, but in Europe at least their aged residents now stay on by choice. In other places, alas, that is not always true.

The Bible tells of many other illnesses. Some sound familiar; thus, Moses threatens his flock that if they do not obey the Lord's commands he will smite them with 'a consumption, and with a fever, and with an inflammation, and with an extreme burning' but nobody has yet identified the Blasting, the Mildew, and the Botch of Egypt that were also used to intimidate those who broke the rules.

Perhaps plague itself – the Black Death, which brought as much terror to the Middle Ages as had leprosy to the people of ancient Israel – was among the promised tormentors. The disease is caused by the bacterium *Yersinia pestis* and is in many ways a mirror image of the bane of Leviticus.

The contrast between the two ancient tormentors could not be greater. Leprosy is – apart from the odd armadillo – restricted to man, is hard to pass on and takes years to kill, while plague is found in many other animals, strikes with great speed, can be lethal within days and is highly contagious. Different as they are, each bears a wider lesson for students of disease; that our enemies – and our defences against them – have evolved, are evolving, and will continue to evolve, whatever medicine does to combat them.

The fifteenth-century Welsh poet Ieuan Gethin, in a diatribe against the fate of his seven children, wrote that 'We see death coming into our midst like black smoke, a plague which cuts off the young, a rootless phantom which has no mercy or fair countenance ... it is seething, terrible, wherever it may come, and a head that gives pain and causes a loud cry, a burden carried under the arms, a painful angry knob, a white lump. Great is its seething, like a burning cinder ... It is a grievous ornament that breaks out in a rash ... the early ornaments of black death.' A century earlier Boccaccio, in the *Decameron*, a tale of a group of young people who flee from Florence to escape that city's epidemic wrote that 'The violence of this disease was such that the sick communicated it to the healthy who came near them, just as a fire catches anything dry or oily near it.'

Bubonic plague, the subject of the Welsh account, is marked by painful lumps or buboes, swollen glands that ooze blood and lymph. The pneumonic form of the disease attacks the lungs until the patient spits blood, while the septicaemic variety infects the blood itself. Each variant is painful, stressful, and in the end often fatal. Many of those afflicted died within a week, and some within a day. In some visitations, almost all those with blood-poisoning perished, and even with the milder bubonic version four out of five did not survive. Again and again, from ancient times onwards, corpses were piled high in cities and villages, so frightful was the carnage.

As Boccaccio noted, its agent is highly infectious. Unlike leprosy, plague enters the human population almost by accident, for it finds its permanent home within various rodents. The bacterium depends on rats and fleas or lice to spread to men and women. It is common among small mammals in central Asia, in a certain marmot most of all, and in the animals is transmitted by contact with infected corpses, or by fleas. Those insects bite other mammals, humans included. As the bugs teem in the fleas' guts, their intestines become blocked and they regurgitate millions of bacteria into the next creature they sample. As a result, many epidemics began in ports such as Bristol, London, or Constantinople, where flea-ridden rats ran ashore.

Pneumonic plague can be picked up from no more than a cough, which means that patients must be kept in isolation (as must their pets, for cats have passed it on to their owners). People can also be infected when they handle contaminated animals or meat (camel meat in central Asia, or guinea pigs in Peru or Ecuador). The illness can spread at great speed, for birds that nest in rodent burrows, or have mice in their own nests, may carry fleas and their passengers for miles. The bacillus can in addition survive in a dormant state in the soil.

The evidence for its presence in ancient Israel is equivocal. The First Book of Samuel speaks of a time when, at a crucial moment in the endless wars of those years, the Philistines seized the Ark of the Covenant and held it hostage. Soon they faced a great epidemic, for the Lord smote their men with 'emerods in their secret parts.' In addition he gifted them with an outbreak of mice.

Alarmed by this double blow the Philistines agreed to restore the Ark to the Israelites, together with reparations of five golden emerods and five golden mice. Perhaps the condition was indeed bubonic plague, the emerods its bloody swellings and the mice in fact rats that passed on the disease. Others suggest that it was instead dysentery, which can produce piles (haemorrhoids, or 'emerods' in the language of King James) and that the mice were just mice, with rats then unknown in the Middle East. Yet others have come up with the idea that the infection was tularaemia, a mouse-borne illness whose symptoms look rather like those of the plague and which could have been brought to the Philistines by mice that nested in the Ark itself. On that object's return to its rightful owners, and as further proof of the power of the Lord (who was outraged that the Israelites had dared to look inside the holy relic) fifty thousand Jewish men died of the same disease, in support of the notion that a creature – perhaps an infected mouse – within the sacred casket was indeed involved.

Whatever the true fate of the Philistines, since their day plague has ravaged the world. Like leprosy it attracts the attention of the pious, who often blame malign intervention, sometimes by Jews (in the fourteenth century, after a visitation of the Black Death, all the Jews of Mainz and Cologne were killed, to no apparent effect). That epidemic reduced the world's population by a quarter.

Its agent's limited genetic diversity makes it possible to search out its roots. The bacillus was born in East Africa, spread from there to the Bible Lands and then moved on to the rest of the Old World. There were two invasions of Asia, the first through India and on to the Philippines, and a separate northern route from the Middle East through Turkey and Iran to China and Japan. Many later attacks in the Levant and in Europe involved a return journey from the Far East, borne by travellers along the Silk Road.

The Plague of Justinian struck the Byzantine Empire in AD 541. It killed half the people of Constantinople and reached across Europe. The next great incursion came in the fourteenth century, again from the Far East. Its passage was helped by the unhygienic habits of a Mongol horde who improved their siege of a Crimean city with a catapult that fired the corpses of those who had died of the disease into the streets (a trick updated in the 1930s by the Japanese, who dropped infected fleas onto Chinese towns). Terrified, the Genoese defenders fled homewards, taking their parasites with them. Around the Mediterranean the epidemic put paid to three in four of the populace. In northern countries its effects were less dire but the illness still led to the death of one in five. A quarter of all German villages disappeared and the economy of the whole continent went into reverse.

The plague came to London in 1348. The then Mayor, Richard de Kislingbury, ordered that two pits be dug, at East and West Smithfield. Within two years, more than a third of the city's residents had succumbed. Thousands were buried in the new cemeteries. The sickness came back again and again, with its last great incursion into these islands in 1665 and 1666, when much of the population of the capital fled. The sole public health measure was a demand that all cats and dogs be destroyed (which was unfortunate as they might have killed the rats that helped spread the disease). The infection continued to flicker in and out of life in Europe and the Ottoman Empire until the mid-nineteenth century. A little later the last global episode – the so-called 'third pandemic' – flared up in China, spread as people escaped from a Muslim rebellion, and crossed the globe. It killed ten million in India alone and lasted until 1959, but by then casualties had dropped to just a couple of hundred a year. The threat was not over, for in 1994 there was a new Indian outbreak. There were no more than seven hundred cases and just fifty-six deaths, but many of the people of the affected region fled in panic. The condition is today classified as a 'newly emergent disease', for it lurks in its homeland, the world of the rodents, and now and again sallies forth to find a few more victims.

The agents of plague and leprosy, like those of malaria, of AIDS and of many other conditions, are in constant competition with their hosts. Disease is a microcosm of the machinery of evolution, for natural selection acts on humans to improve resistance, and the parasite must keep up, or disappear. For all infections, the Darwinian race is on.

Leprosy and its host have almost reached a dead heat. Several variants protect against the bacteria while others control how fast symptoms progress. The Prata leper colony in Brazil, on the edge of the Amazon forest, was founded in the 1920s by Dominican monks and for years was almost cut off from the rest of the country. More than two hundred of its two thousand inhabitants have Hansen's disease – the highest incidence in the world – and almost all descend from earlier generations who faced the same problem. The chances of illness are much influenced by what version of a particular gene, its function unknown, anyone inherits. About one unlucky male in twenty bears two copies of a variant that renders them susceptible and almost all have symptoms by the age of thirty. Those with two copies of the alternative form are more or less safe. A scan of the whole genome in other patients reveals half a dozen other genes that differ in structure between those with a severe, and those with a mild, response to the bacterium. Some match elements of the parasite's cell membrane while others are involved in the immune system.

Plague, too, has forced its victims to change but there the duel between the parties is still in full fury. A variant of a certain receptor molecule on the surface of white blood cells is common in places once ravaged by the Black Death. In the fifteenth century, some Adriatic islands were visited by an incursion that killed seven in ten and they, too, have a higher incidence of that form of the gene than do uninfected islands nearby, which suggests that it gave some protection.

Other mammals, too, evolve when faced with the bacillus. *Yersinia* was introduced into the United States in about 1900 and laboratory stocks of prairie dogs whose ancestors were trapped in Colorado and Texas, where the disease has become common, are much more able to survive infection than are their fellows in South Dakota, where it is unknown. Genes for resistance must be involved. The same is true of black rats in Madagascar. Analogous variants might be present in humans.

Plague and leprosy are at different points in a biological spectrum. Each has made a passage from animals to humans, each was obliged to change as it did so and each is still changing. The new comparative anatomy – the dissection of the double helix – shows how they did it.

Leprosy DNA has been read from end to end. It codes for just three thousand genes, of which no more than half make a protein. The remainder are mere relics of what they once were. The bacillus has lost a quarter of the genes found in its close relative, the agent of tuberculosis.

Plague has also simplified itself, but to a lesser extent, for in the world of the rodents it must switch from host to host to stay in business and cannot concentrate on a single narrow target. It is related to a bacterium transmitted on contaminated foods that does no more than cause stomach upsets. Since it set out on its own the agent of Black Death has changed in many ways. It has lost a substantial proportion of its genes but has gained others that enable it to survive within a flea, to spread through human blood, to extract iron from red blood cells for food, and to avoid its host's immune defences. More remarkable, *Yersinia pestis* has picked up some of the variants that allow it to attack men and women not by mutation but by theft. The genes sit on pieces of mobile DNA called plasmids that have moved in from other bacteria. They are a menace, not just because they donate virulence to a once benign organism, but because some allow their bearers to break down drugs. They can even be passed from harmless bugs to the real killers. In 1995 the first multi-drug plasmid of this kind emerged in a strain of plague in Madagascar. It gave simultaneous resistance to eight antibiotics. The invader resembles another often found in bacteria from the guts of cattle, pigs and turkeys but it has not yet picked up the relevant section of DNA. If the resistant forms spread, plague – now more or less under control – may again become a global threat.

The plague bacillus was for many years divided by the experts into three varieties, the 'antique', the 'medieval' and the 'oriental', each defined by its needs when cultured in the laboratory. They were thought to have given rise in turn to the first, the second, and the third great plagues. The virulence plasmid has, the DNA shows, been incorporated only into the recent, oriental, strain. It is also present in specimens from the nineteenth century and from skeletons from a plague pit in Marseilles filled with remains from an epidemic in 1722. Some of the victims of London's 1348 disaster have now been disinterred, and in a technical *tour de force* the genome of their killer read from end. A comparison of its structure with that of material from three centuries later and of several modern strains hints that both the fourteenth-century event and the

Roman Plague of Justinian also involved that variant. If the bacterium has accumulated new mutations at the same rate as it did between the thirteenth century and the present day, its entry into the human race can be dated at around two and a half thousand years ago, which is just when the Philistines had such trouble with their emerods.

Plague still kills, but if caught early can be cured with antibiotics such as streptomycin. Molecular tests diagnose it in hours rather than the days once needed and allow treatment to start at once. Rat control has also helped. Even so, the illness will never be wiped out, because it finds a home among millions of wild rodents, often in remote places. The pandemics may have gone, but five thousand cases, with more than a hundred deaths, still turn up each year. Most are in Africa.

For leprosy, the future is more hopeful. Some day – like smallpox before it – the biblical scourge may be gone forever. For years, ignorance prevailed. The practice of laying on of hands lasted for millennia and both Elizabeth I and Charlemagne touched the afflicted in the hope that this would act as a cure. Another fad was to bathe in blood, that of a virgin best of all. If even that did not work, castration might be tried. Such treatments were succeeded by a variety of herbal remedies of equal efficacy.

A quarter of a million new cases are still diagnosed each year, but the infection is in retreat. It remains endemic in parts of the tropics but in Europe declined from the fourteenth century onwards, perhaps because of the spread of tuberculosis in crowded cities and the cross-immunity that bacillus gives against its relative. By the eighteenth century the illness had more or less disappeared from England, but survived for a further century in Scotland and until the 1950s in Scandinavia. In the 1930s came dapsone. The drug emerged from research by German chemists on certain dyes. They were, almost by accident, found to kill bacteria. That discovery led to the sulphur drugs, the first successful treatment for blood infections and for a range of other diseases. Dapsone interferes with the bacterium's ability to make DNA and, as a result, reduces inflammation. The drug is not very powerful and for advanced cases must be continued for a lifetime. It can work, but many patients gave up before their medicine had time to do its job. Such rejection was a problem in India. Gandhi himself welcomed a leper, an 'incurable', into his home, an event celebrated on a postage stamp with the message that 'Leprosy is Curable'.

Thirty years ago dapsone was shovelled out in quantity. The bacillus evolved to meet the challenge, resistance emerged, and the drug lost its power.

In 1981 came a breakthrough. A cocktail of three chemicals (dapsone included) was found to be far more effective against Hansen's disease. The ingredients are now supplied free to the World Health Organisation by the drug giant Novartis. They work fast – so much so that, as treatment goes on, the death of billions of bacteria leads to the formation of painful nodules full of dead cells. Some patients abandoned the pills as a result. Then, quite out of the blue, thalidomide – once responsible for birth defects in the children of pregnant women who were given it as a sedative – was found to have almost magical powers to solve the problem. It acts against inflammation, and patients who are in torment with a rash of weeping pustules may show signs of improvement within a day.

Three decades ago, the World Health Organisation set the end of the last millennium as a target to reduce the global incidence of leprosy to a single case in every ten thousand people. It succeeded. Now the aim is to drive the disease down to that frequency within each individual country. That goal is harder to reach but is worth the effort, for success would mean that the agent might then die out of its own accord. In 2010, the WHO recorded a drop by more than half in the number of new cases compared to the figure a decade earlier. Leprosy's capital is now in south-east Asia, but there its agent is in rapid retreat. The Americas are more recalcitrant, with little improvement in the past few years, but global incidence is on its way down, except in failed states such as Sudan and Yemen. Hansen's disease will be around for a while, but a vaccine is under development together with plans for low doses of preventative drugs to be distributed to whole populations at risk. One day the 'unclean' may be no more than a memory.

As plague and leprosy are pushed back the question remains; where did they come from? They are not divine scourges but they do have a history that dates to biblical times and before.

Once again, DNA is the key. It shows, first of all, that each emerged from a lucky (or unlucky) accident. Darwin noticed that the remote islands of the Galapagos had fewer species of plants and animals than did the distant shore of South

America because, by chance, just a few species, and a few individuals of each, made it across the oceans. Island creatures – humans included – are in addition less variable than their mainland relatives. Such bottlenecks have had a dramatic effect on the agents of disease. Many among them entered humankind just once, or on a few occasions. Like birds blown to a remote archipelago, or the first Americans as they pushed into a new continent, most of the travellers fail, but now and again one succeeds in its new home and its descendants explode in number. The genes show that both leprosy and plague arrived as single invaders, each a pioneer of trillions of cells. Their descendants exist in untold multitudes but vary, as a result, scarcely at all.

The leprosy bacillus is, apart from a few minor shifts that have emerged by mutation since it entered its present host, a single clone across the world. Strains from India, from Thailand and from the United States, together with the ancient relics found on the shroud at the Field of Blood, are 99.995 per cent identical. Plague tells the same story. Its agent is almost the same from place to place; a clone, or a few clones, that spread at great speed and picked up the occasional mutation and gene transfer on the way. The strains found among wild rodents are quite a variable bunch, but just a few have made the transit into *Homo sapiens*. The agents of anthrax, of smallpox, of pneumonia, of a certain form of diarrhoea are also of this kind. Syphilis, too, is almost a single clone, and emerged as an inadvertent retaliation to their exploiters by Native Americans, who passed the disease to the Spanish invaders, who took it to Europe in 1495. Its genes show it to be a close relative of a strain of the skin infection yaws that is found only in the New World. A virulent version of a drug-resistant bacterium that causes severe blood-poisoning – methicillin-resistant *Staphylococcus aureus* or MRSA – spread across Europe from Germany in the 1990s. It can kill in weeks. It too is uniform, and its founder cell must have mutated towards virulence within the past couple of decades. Typhoid, in contrast, consists of at least eight distinct lineages, each of which invaded *Homo sapiens* of its own volition. Even so, for many of our enemies the entrance ticket into the human race was won by a fluke.

Where did such creatures live before they found their new home? The Levitical obsession with the cleanliness or otherwise of animals has a basis in fact (even if many of those it names as 'clean' have been a source of infection). Men and women are but islands in an archipelago of illness. Almost all agents of disease have relatives among animals, wild or domestic. Often, the entry into humans coincided with a change in our relationship with such creatures, as farmers began to till the soil, as the first cities emerged, and as people today move into once virgin landscapes and fly in vast numbers across the globe.

Around fifteen hundred different organisms are known to invade the human body. Two out of every three come straight from a wild animal. Few go any further, which means that from their point of view men and women are dead ends (even if they kill their hosts before they themselves expire). The raccoon worm, for example, is more or less harmless to its usual target but should a child eat its eggs (found in faeces) the invader may get to the nervous system and can be lethal. The infection is bad news for both parties.

Other parasites have had more success. They have adapted to their new habitat and, as always in evolution, have made compromises to do so. Leprosy and plague are a microcosm of the expedient and ingenious struggle between ourselves and our biological foes. The battle may end in the death of either parasite or host, or in a disease that allows its victim to survive before its agent moves on. In other cases there emerges an uneasy truce in which both players persist (albeit at some cost to health) and in time such a relationship may even become an alliance from which each signatory gains.

Any intruder must balance virulence – damage – against infectivity, the ability to find a new target. The outcome may depend on how easy it is to find the next victim. A safe refuge in the outside world, a reservoir from which the pool of potential hosts can be topped up, allows an invader to pay less attention to the upkeep of its human home, for should that individual perish before the bacillus can infect somebody else, plenty of its relatives are waiting behind the scenes. The reserve may be within wild creatures, or may consist of resistant spores or particles (such as those found in anthrax or tuberculosis) which can survive in the open for months. All this abets aggression, for such parasites face less pressure to keep their hosts alive than do those, like leprosy, that must move directly from person to person.

The agent of typhoid fever travels from one person to the next in unclean food or polluted water and can sometimes kill, but many of those who pick it up remain healthy but infective, sometimes for a lifetime. 'Typhoid Mary', who worked as a cook in New York in the 1900s even when she was forbidden by the courts to do so, passed on the bacterium to fifty people, three of whom died, before she was confined in a hospital on an island in the East River. Cholera is another water-borne disease with much the same symptoms but is much more lethal. Unlike the agent of typhoid it finds a safe long-term haven in the egg masses of mosquitoes and in tiny freshwater relatives of crabs. It can hence afford to kill off its human hosts without breaking the chain of transmission and often does so. Like plague, cholera owes much of its aggressive behaviour to an incorporated piece of DNA, and is quick to adapt to changes in conditions. In Angola – which had reported no cases since 1998 – a new and lethal variant emerged in 2006. It bore a new alien helper and soon spread through the slums that emerged after civil war forced people to flee to the cities.

The equation between virulence and infectivity is not absolute for some dangerous diseases such as tetanus (caused by a soil bacterium) are not highly contagious while others such as rabies kill off both dogs and humans at great speed, even without a third party to act as a refuge. Even so, each step in the journey from harmless outsider to mortal enemy can still be seen in the world of the parasites. Plenty of animals have diseases of their own that do not bother us. Bird malaria does not attack people and nor does the viral distemper that can kill dogs. Other agents, like the raccoon worm, in effect commit suicide should they enter a human host. Yet others, like plague, do kill, but need a second or even a third host to allow them to persist. For them, no helper means no hope. The West Nile virus illustrates the problem. It causes high fever and is passed on by mosquitoes. It originated in Africa, spread to Europe, and appeared in the United States in about 1999. It may have been carried by a migrant bird, by a feverish human or by an insect stowaway on a long flight. It took just four years to cross the country and did so with the help of introduced Old World mosquitoes. Europe faces a similar problem, for the tiger mosquito, a recent invader from Asia, can bear the agent of dengue and is on the move across Italy, Greece and France. Just one bite of a diseased traveller home from the tropics may be enough to start an epidemic.

Other pathogens move from one victim to the next often enough to spark off a brief wave of infection, but to persist need a regular top-up from an animal host. The many variants of Influenza A (most of which come from pig and duck farms in China) are of this kind. They cause a major outbreak every few years, but die out as their targets expire or become immune. Not until a new variety to which most people are susceptible escapes from its swinish refuge can a new epidemic take hold. The Ebola virus, an African disease that causes patients to vomit blood and can kill in days has taken the next step in the journey by starting to cut out the middle-man. The illness was first noticed in 1976 when it attacked people who had been in contact with infected chimpanzees (and twenty butchers died when they cut up a tasty specimen). Antibodies show that it is not a new disease, for as many as one person in five in some places has been exposed to it over the decades. It can be lethal to gorillas and chimpanzees and an infection from those sources can still kill us, but the virus can now also pass directly from patient to patient in hospitals.

Other afflictions have taken the final step and maintain themselves as exclusive residents in a human host. Smallpox, syphilis, measles, leprosy, the human immunodeficiency virus and tuberculosis are all of this kind. Every infected person must pass on the relevant agent to at least one other individual if the disease is to maintain itself.

Sallies across the species barrier that spark off new infections tend to happen at times of upheaval. Skeletons from before agriculture show that plenty of people had malaria, gut microbial infections and a variety of parasitic worms, although they were free from smallpox and the common cold, which need large populations to survive. These diseases led many to premature death or to a life of misery. With the first farms, men and women began to form new relationships with creatures as different as mice and elephants (the latter among the few domesticated creatures not known to have passed on any agent of infection). At once, the empire of illness extended its borders. Because more animals were tamed in the Old World than in the New, most of its citizens originate there rather than in the Americas.

Measles is related to the cattle disease rinderpest, itself declared extinct not long ago. It found a home in humans with the first herds, seven thousand years before the present and, in spite of vaccination, remains a menace. Tuberculosis, too, is an ancient scourge that resembles a cattle infection. Smallpox arrived three thousand years later – within biblical times – perhaps with a virus that leapt from camels to their riders.

In spite of such attacks, human numbers continued to grow. Soon, the swarms of people imposed a new ecology upon mankind as the first towns and cities (Babylon included) appeared. At once, they became hotbeds of contagion, although New World conurbations such as Tenochtitlan, later to become Mexico City, reached a similar size with no epidemics as they had few domestic animals. A dozen or so of today's diseases must have arisen after such places emerged, for they pass from person to person and would never survive in a scattered rural population. The killers found further opportunities as contacts increased. Enemies travelled faster than did civilisations. A great pestilence in Athens at the time of the Peloponnesian War in 430 BC was described by Thucydides as an invader from Libya via Egypt. It devastated the city ('the bodies of dying men lay one upon another and half-dead creatures reeled about the streets and gathered round all the fountains in their longing for water'). The culprit may have been typhus, passed on by lice and still the companion to war, from the trenches of the Somme to refugee camps in Rwanda.

Imperial China's health records stretch from 243 BC, the time of the First Emperor, to 1911 when the Manchu dynasty collapsed. They tell of almost five hundred outbreaks of illness. For its first three centuries, the empire saw almost no epidemics. From the time of Christ to that of William the Conqueror the numbers of citizens and of outbreaks each increased but most of the populace was spared for decades at a time. In the twelfth century, as numbers grew further and as thousands moved into towns, attacks took place once every couple of years. Most did not spread far. As the population grew yet larger and as people began to move between cities epidemics happened every year. Soon, they covered vast tracts of the empire. Some, such as the incursion of plague in the thirteenth century, killed half the Emperor's subjects.

European explorers much extended the invaders' reach. The people of the New World, of the Pacific and of Australia – until then spared – were devastated by plague, leprosy, tuberculosis, smallpox, malaria and measles. The entry of Captain Cook into Polynesia came at huge cost, for the natives were exposed to enemies against which they had no defence. His diary records the ravages of venereal disease (brought, he assumed, by the French): 'They distinguished it by a name of the same import with rottenness, but of a more extensive signification, and described, in the most pathetic terms, the sufferings of the first victims to its rage, and told us that it caused the hair and the nails to fall off, and the flesh to rot from the bones: that it spread a universal terror and consternation among them, so that the sick were abandoned by their nearest relations, lest the calamity should spread by contagion, and left to perish alone.' Not until the 1960s did the Pacific regain the numbers who had lived upon its islands before the *Discovery* arrived.

The age of infection that began with the first farmers lasted for ten millennia. For most of that time, global life expectancy was no more than twenty-five and most deaths were due to contagions of various kinds. In the developing world they still kill half of those who perish before the age of forty. The parasites that made the leap from animals to ourselves have had a long and comfortable career.

Five decades ago, their reign seemed to be almost at an end. Medicine would soon lay its metaphorical hands on the sick and cure them all. In the 1960s confident claims were made, in books with titles such as *The Evolution and Eradication of Infectious Disease* that 'It seems reasonable to anticipate that within some measurable time, such as 100 years, all the major infections will have disappeared.' In those happy days few remembered the visitations of plague, syphilis, cholera and Spanish flu, each of which had killed millions not long before. Then came AIDS, pessimism and panic. Older enemies have also reminded us of their power. Tuberculosis, once almost gone from developed countries, has returned and now infects more people than ever before. There have been outbreaks of cholera and of swine-flu as further reminders of what may lie in wait. Even trench-fever, a louse-borne illness associated with the horrors of the First World War, has made a come-back among the homeless in the United States.

The price of health is eternal vigilance. Without it, our enemies may again find themselves in the same happy position as Captain Cook when he stepped ashore in Australia, in a rich new continent ready to be exploited. Society is in the midst of a transition as great as that at the foundation of Babylon. Mankind is on the move as never before, across oceans and borders and into jungles, forests and deserts, in a search for food, for minerals, or for suburban comfort. As people move they expose themselves to the dangers that destroyed the empires of long ago.

At least thirty illnesses have emerged from contact between humans and animals in the recent past. They include the human immunodeficiency virus, hepatitis C, bovine spongiform encephalopathy or BSE (which manifests itself in humans as the brain disorder Creutzfeldt-Jakob disease) and several bloody fevers of viral origin.

Apes and monkeys have donated AIDS, hepatitis B, the most severe form of malaria, dengue and yellow fever. HIV made the leap from a chimpanzee to a West African hunter not much more than a century ago, but now thirty million people carry it. It is a disease not just of humans but of chimpanzees themselves, for infected animals die young. Our close relatives have plenty more on offer. Africans who hunt or butcher wild apes and monkeys, or those who keep them as pets, are most at risk. In Cameroon – where HIV first moved from chimps to humans – monkeys and apes carry infections from another group, the simian lymphotropic viruses, which enter white blood cells. Many hunters bear them, as do members of their families, for, like HIV, the particles are transmitted during sex. The virus causes muscle damage. A related pathogen, the chimpanzee simian foamy virus, is also common among the citizens of Cameroon, but in spite of its close resemblance to the agent of AIDS it seems to do no harm.

Pigs, too, are threatening creatures. Not only are they the source of swine-flu, they can also carry the Ebola virus, which may one day leap from that widespread reservoir to ourselves. Twenty years ago in Malaysia piggeries were set up close to fragments of tropical forest. There, in 1998, the Nipah virus crossed from fruit-bats to pigs and thence to those who farmed them. It killed a hundred people. A million animals were slaughtered. That did not hold back Nipah, which moved as far as Bangladesh. Once transmitted from bat to pig, and from pig to man, Nipah has now gained the ability to pass from person to person with no help from a third party.

Bats bear many other lethal viruses, rabies and Ebola included. Every fifth species of mammal is a bat, and the destruction of forests has encouraged the animals to move closer to towns and cities. The outbreak of severe acute respiratory syndrome or SARS that began in China in 2003 and crossed the world was first blamed on civet cats sold for food. It was contained with quarantine and with the slaughter of thousands of wild civets but now it appears that, once again, bats are to blame. Fruit-bats have been known to fly three thousand kilometres. They might as a result carry once rare illnesses to great cities such as Sydney, which is, perhaps mistakenly, proud of the colony in its Botanic Garden.

Other parasites have leapt into action as people themselves feel the call of the wild. Lyme disease, which leads to a painful inflammation, is caused by a relative of the agent of syphilis. As suburban homes sprang up in the northern woodlands of the United States, the locals came into contact with deer and mice and their infected ticks, and suffered as a result.

Not all the news is bad. For leprosy, plague, polio and more there has been real progress. Many other illnesses – malaria, dengue, and cholera included – could also be mastered if the political will was there. Smallpox has disappeared, polio may be on the way out (even if there has been a resurgence since some Islamic authorities declared vaccination to be a plot to sterilise Muslims) and the last days of leprosy may not be far away. The parasitic worms of Africa have been driven back at speed and optimists claim that many of the other epidemics that began with agriculture may be defeated in the next few decades.

Another reason to be cheerful is that some (but not all) dangerous illnesses appear to moderate their effects as the years, and the centuries, go by. Smallpox became less virulent in the years before its elimination, while the 'great pox' (otherwise known as *Morbus Gallicus*, the French disease, even if the French blamed it, correctly, on Native Americans) that swept across Europe in the Middle Ages and killed in a few months, has slowed its progress in its modern form, syphilis. Cholera, too, is in most places less lethal than once it was. The viral illness dengue, or break-bone fever, in contrast, has travelled in the opposite direction. Until the 1950s, almost all those affected had just a single infection, which was painful and unpleasant, but primed their immune system against later attacks. Now, other variants have spread with the global movement of population and a second infection with one of those may be lethal. The evolution of virulence, like that of everything else, does not follow simple rules and the claim that all diseases follow a programmed path to be kind to their hosts is not correct.

Even so, some of our erstwhile enemies have moved towards uneasy compromise. Other disease organisms have gone further. They have moved so close to an accommodation with those who bear them that, most of the time, they are not noticed. Like the benign skin fungi that attack AIDS patients only in the last stages of their illness, or the bacteria that do their damage when the body is weakened after surgery, they are a risk only to the susceptible, be they the aged, the frail, or those with a damaged immune system. In the same way, a certain skin bacterium causes no problems until hormonal changes at puberty allow it to invade glands and cause acne.

Some of our hangers-on have taken yet another step for they live with us not as enemies but as at least temporary friends.

Each reader of this book is less human than they were on the day when they were born. Every baby emerges pure into the world but, within minutes, is invaded by battalions of foreign cells that soon teem throughout its body. All adults have ten times as many bacteria in their guts as they have cells of their own (which means, in terms of cell count, that the equivalent of just one leg below the knee is truly human). Billions more are secreted around the nooks and crannies of the skin, the mouth and elsewhere. Each of us is laden down by more than a kilogram's worth of these tiny outsiders. Together, they weigh as much as the brain and there may be ten thousand distinct kinds, each with its own set of inherited information, which means that our residents contain a far greater diversity of genes than do our native cells. Everyone's mixture is as unique as a fingerprint (and the police have used that fact to see who has been at work on a particular computer). Its identity changes little from day to day or week to week although it does shift in old age. The forehead's community is distinct from that of the arm; and the leg from the sole of the foot. Mouth, stomach and vagina each have fewer types than do other sites (and, for some reason, the back of the knee is more diverse than anywhere else). The mix also varies in identity across the world.

Such aliens were once seen as harmful, and a whole generation of cereals and laxatives promoted 'inner cleanliness' by expelling them. The habit goes back to the Egyptians who, three thousand years ago, saw faeces as a cause of disease and gave themselves enemas every few days. More recently, many doctors claimed that to get rid of the noxious hangers-on would avoid 'auto-intoxication' because of their presumed tendency to poison their carriers. Some people thought that they were responsible for age itself and told surgeons to cut out bits of intestine in the hope that they would live longer. The Nobel Prize-winning immunologist Élie Metchnikoff was convinced of the dangers of such 'putrefactive microbes' and dismissed the entire large intestine as a vestigial organ of no use. On his death-bed at the age of seventy-one, he asked the doctor who was to do his post-mortem to pay particular attention to the length of his gut in the hope that there might lie the key to his success in reaching three score and ten.

The notion that the inner self is unclean is quite wrong. Instead its denizens are a forgotten organ, as essential as the kidney or the liver. We depend as much on our internal citizens as they do upon us. In the large intestine, they break down indigestible plant tissues to feed themselves, but as a by-product they make chemicals that can be soaked up by our own guts. In addition they neutralise poisons and get in the way of pathogens that might otherwise settle in. Their biological signals help gut muscles and nerves to develop, which means that children who possess a certain combination of gut residents are at higher risk of obesity and of diabetes than are those with a different mix.

Should the internal ecosystem be disturbed by an attack of aggressive bacteria, the appendix – once assumed to have no useful role – acts as a reservoir from which the useful cells can emerge when the threat is over. Sometimes, the gut flora is wiped out by antibiotics. Other, more noxious, bacteria may then take over and can kill. A simple, startling and often effective treatment is to carry out a faecal transplant; to insert the intestinal contents of a healthy person into the patient in the hope that this will restore the internal balance. The commercial yoghurts that claim to do the same job are, however, useless.

A long history of cooperation has led to an interaction with our inner army so intimate that its absence, rather than its presence, leads to illness. Laboratory mice raised in germ-free conditions do not thrive. Their immune systems do not work well, their intestines fail to move as they should, and they tend to be unhealthy. Such animals are also anxious and obsessive and have a poor memory. The inhabitants of the intestine may once again be involved. They make plenty of the nerve transmitters associated with mood and anxiety. They also generate nitric oxide, which helps to pass information between nerves and is important in emotion. Their unexpected ability to keep us cheerful adds fresh significance to the biblical 'bowels of compassion'.

One of the most important tasks of the colonists is to prime the immune system. The developed world is in the midst of an attack of allergies, of childhood asthma, of skin eruptions and of other conditions in which the body's defences fail to respond as they should. Such conditions include multiple sclerosis, juvenile onset diabetes and irritable bowel syndrome, in all of which the protective protein attacks the body's own tissues rather than the threats from outside.

Some of this may be due to today's excessive cleanliness. Too much hygiene might be harmful because it destroys our internal ecosystem. Parts of the modern world have almost attained the purity hankered after by the Levites. The gut ecology of African children differs from that of Europeans and – although many other variables are no doubt involved – they do not have as many allergies. Middle-class British children from small families without a cat or dog, with homes in cities rather than in the mire of the country, and those who stay with their mothers rather than going to nursery, are safer from parasites and infection than are others and they too suffer more from such failures of the immune system. Turkish immigrants in Germany who often visit their native land have more gastric upsets than do their fellows who stay in Europe, but they also experience fewer immune problems. In the same way, mumps, chickenpox, measles and other childhood diseases are less common among children in poor tropical countries who are infected with hookworms than among their uninfected peers, as a hint that their parasites have improved the efficiency of the immune system. Inflammatory bowel disease – once rare but now common – first emerged as a real threat in rich people living in cold climates, where worms are rare; and the last American group to begin to suffer from it were poor African-Americans.

Twenty years ago, a certain bacillus was found to be the cause of most stomach ulcers. The response was to hit it with antibiotics. The treatment worked. The organism is, or was, found in almost every stomach on Earth. Such is the modern use of antibiotics, with most American children having a dozen rounds of the stuff before they are eighteen (and their British equivalents facing almost as many) that only one in twenty now bears the bug. That is unfortunate, for the same bacterium has the unexpected task of interacting with appetite hormones. Farmers found some years ago that antibiotics given to intensively reared chickens, pigs or cows makes them put on weight and lay down fat, for reasons then obscure. They do the job, we now know, because they kill off that stomach bug and, in turn, push up the appetite hormones and make the animals hungrier than they were. Good for profits as that might be, it hints that at least part of the wave of obesity that afflicts the developed world may be due to the loss of a once universal stomach bacterium. It is a friend to the young, but an enemy to the old.

For most of history, all infants have had such mild, and not so mild, hangers-on. Fifty million children are still treated for parasitic worms each year. Invasions of this kind involve a conversation between parasite and host as each adapts to the other until, in time, the immune system is sufficiently well educated to do its adult job. Certain viruses, such as that for hepatitis A (a food- and water-borne disease that has little effect on most people) are particularly effective. Antibodies against it were present in almost all British children fifty years ago but the proportion has now dropped to just one in four. The immune system is learning less than once it did.

So strong is the tie between cleanliness and inflammation that a little judicious dirt – or even a well-chosen parasite or two – may help to fend off some of the problems that come from immune failure. Patients with multiple sclerosis sometimes find that a dose of worms reduces the frequency at which the symptoms recur. A refreshing drink based on the live eggs of an intestinal worm of pigs is also effective in around half of all patients with inflammatory bowel disease. A revulsion against that idea has led to the search for the parasites' own chemical signals of identity in the hope that they might prime the immune system instead. Clinical trials are under way.

To treat plagues with human waste (let alone with the denizens of pig guts) would no doubt have shocked the Israelites. Even so, now that the biblical ideal of purity, in body if not in mind, has in part been achieved, the result has been a wave of illness that may be countered by a partial return to the unclean world in which they and their fellows lived. Cleanliness is not as close to godliness as the Good Book makes out. The urge to reject those cursed by infection has been replaced by the idea that impurity has a part to play in keeping ourselves pure. How would Leviticus, the scourge of lepers, have coped with that?

CHAPTER EIGHT

ZWINGLI'S SAUSAGES

William Blake, *Children Round a Fire*

When thou hast enough, remember the time of hunger.

ECCLESIASTICUS 18: 25

Food is everywhere in the Bible. From the Forbidden Fruit to the Last Supper and from the Manna in the Desert to the Feeding of the Five Thousand the Good Book is much concerned with diet. It is set in a land of milk and honey but one also faced with famine; a place of feast and fast, of drunkenness and self-denial, and of great showers of bread from the skies and the transformation of water into wine. Sacrificial banquets, with bread, oil, alcohol and meat are offered to the populace, with slices reserved for the priesthood and the choicest cuts saved for the deity. Women do the cooking. Many are honoured with culinary names; that of Rachel, mother of Joseph, means 'ewe' and the title of Leah, matriarch of six of the Twelve Tribes, can be translated as 'cow'. Judaism itself finds its roots in the fields, for after the expulsion from the happy land in which food could be gathered from the trees Adam was condemned to till the ground. Hunter-gatherers in their modern Edens still eat less starch (bread included) than do farmers (and far less than do devotees of junk food).

Man's equivocal relationship with what he puts in his mouth is older than Judaism, the Fall of Man, or faith itself. *Homo sapiens* is the dyspeptic ape. Compared to our primate relatives we are gutless, with an intestine just half the length of theirs. We have small mouths, reduced teeth, weak jaws and a modest stomach. Our tastes are decayed and in a bland world of plenty we have lost many of the safety mechanisms that control what we swallow. Surrounded as they are by plants whose main desire is not to be eaten, other apes have no choice but to be choosy. The Chimpanzee Restaurant would never get a Michelin star. Its menu sets stiff challenges to the diner. Many tropical fruits, luscious as they appear, have less sugar than a carrot. Some ape staples taste of mustard oil while others cause a human tongue to freeze as if at the dentist. To cope with such meagre fare its hairy customers open their mouths twice as wide as we do and are obliged constantly to stuff in leaves, fruits and the occasional hunk of monkey (and even chimpanzee) flesh. Each day the animals eat twice as much in relation to body weight as does a typical man or woman. That calls for chewing. Humans spend little more than sixty minutes a day at the pastime (which leaves lots of time for other hobbies) while chimps grind their teeth for half the daylight hours.

We get away with decayed tastes and feeble guts because of a unique talent: cooking. A saucepan is an external stomach. With its help *Homo sapiens* has become the ape that bakes, barbecues, bastes, blanches, boils, brazes, brews, broils, browns and (sometimes) burns. Boiling and frying, as Mrs Beeton put it, 'render mastication easy', and her French predecessor Brillat-Savarin, the self-styled 'philosopher in the kitchen', went so far as to claim that 'It is by fire that Man has tamed Nature itself.'

In his 1825 book *The Physiology of Taste,* the French gourmet discusses the nature and social import of cuisine and the origin of likes and dislikes when at table. He pointed out that: 'A man does not live on what he eats ... but on what he digests.' He was right, and his own sometimes indigestible pages on the inner significance of diet have developed from a philosophy into a science. Food has become an art, a craft, a business, and a statement of personal, national and religious identity.

Whatever higher significance might be attached to the kitchen, cooking ensures that its raw material is in part broken down before it gets into the mouth. The process is very effective. Some unfortunates have, for medical reasons, a hole in the wall of the small intestine which allows meals to be sampled on their way through the system. If such people eat raw starch, about half is broken down before it reaches the point where it can be absorbed. Boil the stuff first, and ten times as much is digested by the time it enters that section of the gut. The chemical changes brought about by heat double the amount of useful energy provided by certain vegetables and by almost as much for fried eggs. An item not passed over fire obliges the body to use its own fuel to process it, so that it takes more work to extract its ingredients. Heat is useful in other ways. Many plants, from parsnips to kidney beans, have poisons that can be destroyed by a fiery blast. The stove (or the microwave) kills bacteria and ensures that our immune system (which demands lots of energy) has to leap into action less often than it did in less refined eras. Even better, some prepared foods last longer than do raw. Most important, the urge to cook means that we spend much less time at the table and hence have the chance to do more useful things.

The kitchen made us what we are; as Boswell put it: 'Man is a cooking animal . . . no beast is a cook.' He was right. Fire gave our ancestors food for thought. It freed them from the drudgery of digestion and allowed them to grow huge brains on the cheap. The contents of the human skull are three times bigger, but the intestine a quarter shorter, than those of our closest relative. Our brain cells are no bigger than those of chimps but we have far more of them, which explains why an organ that weighs a fiftieth of the body mass takes a fifth of the energy budget (and in small children even more) while the chimpanzee invests only a third as much in cogitation. The enzymes that break up glucose, the fuel of the body machine, are more active in human brain than human brawn, but in chimpanzees the balance goes the other way. Men and women have expensive, large and efficient brains but cheap, reduced and ineffective guts with which to provision them. The saucepan squares that energetic circle.

We have become slaves to the stove. Man cannot live by raw food alone, although plenty of people have tried. Marco Polo claimed that Mongol warriors of his era could live for a week on nothing but horse blood, and their intellectual descendant Arnold Schwarzenegger in his muscular days swallowed raw eggs mixed with cream for breakfast (the recommended dose was three dozen but had he boiled them first he could have managed with fewer). Some obsessives, the 'rawists' as they call themselves, choose uncooked food for ideological reasons. They are a schismatic band. Fruitarians vie with juicearians, sproutarians, alkalinists, palaeolists (those who demand foods available in the Stone Age) and anopsologists (disciples of the diet they imagine to have been eaten by our pre-human ancestors) in the purity of their devotion to what they see as the most natural or the most moral cuisine.

Whatever their quarrels about the details, those who abjure the kitchen are united by a shared and painful sensation: hunger. It reminds them of what existence was like for almost everyone for much of the past and for some people today. A diet based on unprocessed food alone cannot long sustain life. It lacks the essential amino acids freed by heat and is also short of energy. Those who try raw meat, milk or fish do rather better than rawist vegetarians but they move from the clarity of their convictions when they accept, as some do, chopped, pickled, fermented, or oven-dried delicacies, for these are in effect partially pre-digested. Even so, in the end they too begin to starve. The BBC once persuaded a dozen sufferers from high blood pressure to eat like chimps at Paignton Zoo. Cucumbers did wonders for their medical problem, but even as they stuffed themselves the experimental apes lost weight and in time would have faced osteoporosis, an increase in 'bad cholesterol' and a lifetime of poor health. All rawists face that difficulty and many of the women among them cease to ovulate (which, in evolutionary terms, is not good news).

The plight of the great uncooked shows how much *Homo sapiens* has come to depend on fire. Anthropologists have a fatal tendency to decide what made us human – we are, they say in their various ways, the upright, the grasping, the naked (or the well-dressed) ape; the handy, the thoughtful, the babbling or the dishonest primate. Those attributes are no doubt important, but the kitchen has been as central to human origins as were any of them. From Stone Age to Aga Stove gastronomy has been a social adhesive. From its first days, life's basic chore became a shared set of talents rather than an individual action. Our thick-browed forebear *Homo erectus,* who flourished from some two million years ago, had already evolved small jaws and teeth and rather large brains, as had the Neanderthals, whose unbrushed fossil teeth still bear the remnants of cooked grains. All this hints that even our distant predecessors spent less time around the table than did their primate kin.

Burned bones almost as old as the first *Homo erectus* have

been unearthed in South Africa but whether the fires were natural or otherwise is not clear. Later relics have turned up thirty metres inside the entrance of the Wonderwerk Cave in the Northern Cape Province. Nearby are the stone lids of pots. The million-year-old flames must have been lit by our ancestors. The first evidence of community life comes from groups of ancient hearths from around that time, as a hint that society emerged as men (and no doubt women) huddled around a barbecue. In spite of the cold, fire in Europe does not appear until half a million years later.

From barbecue to bistro diet bonds diners together. Food is a statement of friendship, of order and of social position. It acts also as a badge of membership of one society rather than another, as all Krauts, Frogs, and Rosbifs are aware.

The Church noticed that fact long ago. Brillat-Savarin points out that 'In Paradise there were neither cooks nor confectioners.' He may have been right, but after the Fall of Man those professions became central to ecclesiastical practice. The Bible tells us that good and evil – and sex and death – began with a forbidden snack, while Christianity's central episode, the Crucifixion, was preceded by a formal meal. An obsession with food has split Jews from infidels, Christians from Jews and the Roman Church from its reformers.

Most creeds have dietary rules that bind their members together. They use them as a symbol of identity. Hinduism in its first days, some versions of Islam, and (in ritual form) both Judaism and Christianity offer sacrifices to their deities to seal their bonds with their subjects. Often, the cost of an entry ticket to a particular system of belief is to shun certain items altogether or to obey rigid laws about how they should be prepared. The Jains, an Indian sect that views all life as sacred, have such a horror of flesh that some wear masks to avoid the risk of breathing in insects. Tibetan Buddhists and Somali Islamists detest fish while Hindus avert their eyes from beef. At the other extreme, the Mayas of Central America had it that humans were created to nourish the gods. The priests smeared the blood of their victims onto the mouths of stone images to feed the deities within. Ten millennia earlier, the people of the Pyrenees sacrificed deer deep within caves and hammered splinters of their bones into cracks in the walls, perhaps as sustenance for a divine presence behind the rock face.

Judaism is precise in its culinary instructions. In *The History of the Decline and Fall of the Roman Empire* Gibbon refers to the 'peculiar distinction of days, of meat, and a variety of trivial though burdensome observances' set out in Leviticus, the book that made the lives of lepers miserable. The manuscript was written in Babylon. There, nostalgia for their native land led the exiles to protect their identity with customs that distinguished them from their captors. On the return to Jerusalem such precepts were taken up by, or imposed upon, the whole population in a further attempt to emphasise the importance of divine ordinance to human activities. The dietary guidelines set out in the Levites' instructive volume are oppressive indeed. They led to Kashrut, the rituals behind the choice and preparation of food (which at the extreme involves the dissection of a particular nerve from the leg of a cow or sheep). The book gives instructions about slaughter to ensure that the chosen people do not drink blood, and defines an eclectic mix of creatures as unfit to eat. Camels, rabbits, pigs and more are prohibited as they chew the cud but do not divide the hoof. Also forbidden are all shellfish (and they are an 'abomination' rather than just unclean, a disgrace shared by eagles, vultures, ospreys, owls and pelicans). Locusts and beetles are edible but other insects are not. Also banned from the table are ferrets, chameleons, snails and moles. Any utensil that comes into contact with such a creature must be destroyed.

The kitchen itself is under strict control. Many Jewish families keep separate plates for dishes of meat and of milk. For the Passover feast – the anniversary of the Exodus – even table manners are prescribed: 'with your loins girded, your shoes on your feet, and your staff in your hand; and ye shall eat it in haste'. Some interpret the verses to mean that the feast must be consumed in Jerusalem. Samaritans construe them instead as directed at Mount Gerizim, near the now tense West Bank town of Nablus. I once attended the Passover ceremony there and gained a grim satisfaction at the gasps of the tourists as they watched the sheep's throats being sliced in the approved way.

The Acts of the Apostles released Christians from such restrictions. St Peter saw 'heaven opened, and a certain vessel descending upon him, as it had been a great sheet knit at the four corners, and let down to the earth'. The divine blanket was filled with all manner of animals and birds, and he was enjoined by a voice to eat them. When Peter objected on the grounds that some were unclean or 'common' the Lord made the helpful comment that 'What God hath cleansed, that call not thou common.' That opened the kitchen door to pork, to shellfish, to moles, ferrets and more, not all of which made it to the table. Even so, the ancient fixation with diet lives on in Christianity in a different way, in the transmutation of bread and wine into the flesh and blood of Jesus. The altar, where this ritual – the Eucharist – takes place is the focal point of many churches. Its significance is imbued into every Catholic child at its mother's knee.

Choices about what to put on the plate may depend on belief and on culture, but biology also plays a part. The decisions reached are of interest not just to those who wish to assert fidelity to a sect, but to the marketers who manipulate what we eat with new, seductive, and expensive items.

There was once no accounting for taste. That era has gone, thanks to molecular gastronomy, the science of culinary experience. Its interests stretch from shelf to sewage works and its successes mean that food has become an even more precise badge of identity than once it was. From China, where meat is a statement of membership of the new middle class, to Britain where the bourgeoisie snack on raw fish while the proletariat prefer Indian takeaways, a vast industry is at work to persuade people to change their habits. Culinary sensations have begun to make sense. The choice of food emerges from a host of evolutionary, social and historical pressures.

The taste buds are fine protrusions on the surface of the tongue, the palate, and the back of the mouth (the claim that those for distinct flavours are in different places is a myth). Individual buds are specialised to perceive food as sweet, sour, bitter or salty. There may also be separate receptors for metallic tastes and for fat.

A century ago a Japanese scientist discovered a new savour in an extract of edible seaweed. He gave it the name 'umami', which means 'tasty'. The sensation is shared by those who eat cooked meat (bacon most of all), shrimp, mushrooms and mature cheeses such as Parmesan. The Romans were addicted to the stuff for they drenched their plates in a sauce made of mackerel guts, which sounds repellent but in fact was rich in umami. The substance is mouth-watering and also speeds up the passage of a meal through the stomach (the Romans used it to relieve constipation). Umami is tasteless until food is cooked or allowed to mature for it binds to proteins and is not released until they have been broken down by fermentation, ripening or heat. Its presence provides the mystic savour of a good stew described by Brillat-Savarin as 'osmazome'.

Its Japanese discoverer set up a business. Its product was the infamous monosodium glutamate, now much used in packaged foods. Some people avoid it on the spurious grounds that it causes 'Chinese restaurant syndrome' with its supposed headaches and nose-bleeds. So widespread is this conviction that the labels of many products claim 'no added MSG' (although they fail to point out that the soya protein used instead is packed with the chemical).

Most people have around five thousand taste buds. Those for bitterness, sweetness and umami bear specialised proteins to which sugars and the osmazome flavour bind. Taste and smell – together with mere anticipation – help prepare the gut for its task. They release digestive enzymes, cause the intestine to ready itself for the work to come and persuade the body to pump out insulin in readiness for a surge of sugar.

Just three genes are involved in the perception of sweetness and umami, with their products combined in different combinations. Foodstuffs with subtly different flavours – sugar versus grape-juice, for example – attach to different points on the receptor. Certain fruits such as the African 'miracle berry' produce a protein that binds to the sweetness receptor and alters its shape so that it responds to acids instead. As a result, bitter substances such as lemon taste sweet until the protein is washed away (an attempt to market the stuff to block the a craving for sweet food was quashed by the American sugar industry).

The proteins involved span the cell membrane like a bent hairpin, with seven back-and-forth traverses of the boundary wall to form a barrel-shaped channel that links the inside to the outside. Each end bears three loops. The outer loops bind the tasty item and alter the shape of the channel. The inner loops then move, and make a chemical signal that is transformed into electrical impulses to the brain.

Dogs have fewer taste sensors than we do and cats fewer again. All the latter's relatives, tigers included, have a damaged sweetness gene, and are not tempted by syrupy snacks. Other carnivores, from hyenas to sea lions and dolphins, have also lost those receptors as proof that to maintain them when faced with a sugar-free diet must be expensive. Pandas, vegetarian as they are, have given up umami. The most talented tasters are goldfish, denizens of muddy water that have sensors all over their bodies and use them to find food.

Bitter substances can be sensed at concentrations a thousand times lower than sweet, perhaps because they warn of poison. Two dozen receptors, some of which can respond to fifty chemicals, are involved. The system has evolved fast in primates because of their history as vegetarians surrounded by items whose concern is not to be eaten. Some plant constituents cause immediate damage, while others, ingested over long periods, harm the thyroid gland. A few substances are deeply unpleasant to some people but tasteless to others. One has long been dropped into beer or tea by genetics students amused by the bafflement of those who can, and those who cannot, sense its presence. A third of all Europeans cannot taste the stuff at all, but one in five is a 'super-taster', able to pick up its extreme bitterness even at low levels. Non-tasters have two copies of a damaged version of the relevant gene. In laboratory tests at least, super-tasters tend to avoid foods such as broccoli and spinach. For them lemon is more acerbic, cream more unctuous and salt has more savour. They are also more sensitive to the sweetness of sugar and to the burn of a curry. More important, such people smoke and drink less than others, perhaps because they can sense the unpleasant tang of the narcotics involved.

Smell is central to the gourmet's experience. To pinch the nose while eating a steak proves its importance for the meat then loses much of its flavour. Scent is subtle. That of the tropical durian fruit is repulsive when sniffed, but delightful when it enters the nose from the mouth. The body uses combinations of receptors to sniff out odours, for one receptor binds a number of different chemicals, while the same chemical may bind to several receptors. It can even distinguish molecules that are identical except for their shape. Caraway and spearmint, for example, differ only in that they are mirror images of each other. Even so, our abilities are much diminished compared to those of many other mammals. The family of genes involved has about nine hundred members but around half are decayed and useless. Individual differences come from changes in the structure and number of receptors themselves, with some people blessed with nine copies of a particular version while others lack it altogether. As a result, everyone inhabits their own odorant universe. For some reason, Europeans have a feebler sense of smell than do people from the rest of the world.

Temperature is also part of the equation, for certain items are acceptable only when hot and others only when cold. Warm water tastes different from chilled, and to heat up a small patch of the tongue without any food at all gives rise to a phantom flavour. Pain, too, finds a place, for the burn of chilli pepper uses the same receptors as those sensitive to tear gas. Its mirror-image, the gentle sensation of mint, comes when such structures are blocked. Touch (what marketers call 'mouth-feel') also plays a part. Toffee and caramel are almost the same stuff, but their distinct textures make them taste different. Appearance is also involved (as Mark Twain put it, 'If you have to eat a frog, don't look at it for too long') and most people will not accept meat dyed blue. Beef is sometimes sold in sealed packages with an atmosphere of carbon monoxide, which combines with the blood pigment haemoglobin to give the item the cheery hue of a baby asphyxiated by a faulty gas boiler. Even the container makes a difference: a soft drink in a blue tin is more thirst-quenching than the same liquid in a red one. Sound, too, gets a look-in; think 'Snap, Crackle, Pop!' (or, if you are German, 'Knisper, Knasper, Knusper!'). The decision as to what we sample, spit out or swallow is more complicated than many people realise.

The mouth is no more than a gateway to the metabolic factory. Its machinery has quirks of its own. Seven out of ten of the world's adults cannot digest milk, while millions more cannot handle alcohol, gluten or excess starch. A nasty experience with one of those can put someone off for life. Such differences have evolved since the origin of agriculture, ten thousand years ago. The gene that gives adults the ability to drink milk is common in Northern Europe, with its many cattle, but rare in most of Africa, where agriculture arrived late. It has emerged there only in peoples such as the Fulani of Nigeria who have cows. In much the same way Northern Europeans who grow corn can digest starchy food better than can the happy tropic-dwellers who pluck sugary fruit from the trees.

Variation of this kind is important in medicine. Around half the people of the globe find it hard to break down drugs such as those used against tuberculosis, and doctors must adjust the dose to fit. The ghosts of poisons past are to blame, for the feeble versions of the relevant genes are far more common in Europeans than in Africans. The latter remained hunter-gatherers until not long ago and had to deal with many plant poisons. They retained enzymes that neutralise them and they do the same to today's medicaments. The bland diet of those who tilled the fields meant that this system was no longer useful and faded away.

The proteins involved in taste and smell are related in shape and size to a variety of other receptors and to hormones that alter mood. Biology is important to the experience of what is on the plate, but habit, custom, culture and emotion also play a part. Brillat-Savarin saw as much. A good meal was aided by the presence of 'a pretty woman, her napkin placed under her arms', of 'conversation steered clear of political arguments, which are hindersome to both ingestion and digestion' and by bowls – not cups – of coffee engulfed 'with a noise that would do honour to sperm whales before a storm'. At the end of a repast, he wrote, both body and soul should enjoy a special form of contentment. Today's chefs follow his advice. An expensive restaurant has fine china, white linen, polite staff and decorous silence. A burger joint does not.

Experience plays a part in many ways. An eccentric experiment in Wyoming set out to keep coyotes away from sheep. Dead lambs laced with a vomit-inducing drug were scattered across the landscape and the predators soon learned to associate their taste with nausea, so much so that they ran away when a sheep approached. People, too, can develop a hatred of foods that make them feel ill. Cancer chemotherapy can cause real discomfort and some patients find that they can no longer abide an item that they had enjoyed just before the treatment. Some doctors now offer novel tastes (such as an unusual ice cream) before therapy in the hope that the aversion will be directed towards that instead.

As Proust realised when the savour of a madeleine projected him back to childhood, memories of infancy often decide what is allowed onto the plate. All parents provide their children with strict instructions on what can and cannot pass their lips. As a result, plenty of items – grubs in Australia, dogs in Korea, snails in France and locusts in ancient Judea – are palatable to some nations but repellent to others. The differences can be subtle, for porridge is sweet in England, but salty in Scotland. Familiarity is at the head of every table; as Mark Twain said, 'It was a brave man who first ate an oyster' (or for that matter a prawn curry). Memories span the generations. Many immigrants to the United States have lost their native language or faith but continue to prefer the food of their great-grandparents, from pasta to schnitzel (and the first English colonists starved when they refused to eat maize offered to them by the locals and insisted on growing European corn instead).

The lessons start young. Kittens who watch their mother eat a banana under the influence of electrical stimulation of the brain are willing to try that item themselves, even if they cannot perceive its sweetness. In much the same way babies whose mothers ate aniseed when pregnant are happier to accept food with a dash of that herb. That predisposes a newborn to prefer familiar (and hence presumably safe) food. Such experiences can be dangerous, for to expose rats to alcohol before birth increases the chances that they will accept the poison when they are older. This may be another reason for pregnant women to avoid the stuff.

For food, as for many other things, the child is father of the man. The best predictor of an infant's preferences when he is eight is what he enjoyed when he was four. Children like what they know and eat what they like. In their earliest days they are content with breast milk, which is sweet, creamy and well provided with umami. Those preferences live on into adulthood.

For most of history they could not be satisfied, but things have changed. What were luxuries have become staples. Junk food is a synthetic version of mother's milk, filled with sugar, fat, and osmazome. That may be why it is so addictive. In earlier times, the poor ate little meat or cream but used grains and beans to fill themselves up. In the fifteenth century, England imported around fifty tons of sugar a year – a fraction of an ounce per person (and most people saw none of it). Slavery changed all that, and sugar became the first mass-marketed product, piled into every Englishman's tea, 'that blood-sweetened beverage' as the poet and abolitionist Robert Southey called it. The average American now eats two tons of the stuff in his lifetime – an increase of ten thousand times in five centuries. A typical child is encouraged to do so, for he or she watches that number of television food commercials each year, almost all of which are for sugared cereals, sweets and soft drinks. In Britain, the 2012 Games – the Obesity Olympics – were, with a certain irony, sponsored by Coca-Cola and McDonald's.

One person in five, worldwide, is overweight and almost half that group is obese. Since the 1960s a wave of lard has crashed upon these shores. A quarter of all British children weigh more than they should, and almost all had reached that state by the time they were five. Across the Atlantic matters are even worse. One in four American adolescents shows signs of 'adult-onset' diabetes, a figure that has doubled in the last twenty years. The majority of the nation's health budget is now spent on metabolic conditions such as diabetes or heart disease, many of which can be traced to diet.

In Victorian England, where obesity was blamed on sloth and indulgence, it was rare but in some places fat has become the norm. On certain Pacific islands half the population falls into that category, and in many American states the figure is one in three adults. In Europe, Norway is the slimmest nation, and Hungary the fattest. Everywhere, women are larger than men.

Swollen bellies are a hangover from the past, for our ancestors evolved in times when shortage was more familiar than glut. The juggernaut of greed has, as a result, a powerful accelerator but feeble brakes. When food is plentiful the body is happier to ramp up the appetite than to counsel moderation, and it finds it harder to shed pounds than to put them on.

Excess is now more of a danger than deficiency. Such luxury is new, for throughout history famine ('great dearth throughout all the world') was common. Hunger is a torment both to body and soul. In the twentieth century seventy million died from starvation. Increased numbers, food price speculation and the loss of land to houses or to factories mean that the biblical scourge is set to come back. Soil degradation alone means that Africa will be able to sustain no more than half its people within the next decade. Across the world, a billion go hungry and many more may soon be forced to join them.

The body's economy, like all others, is ruled by the balance of supply and demand. The average Briton consumes around fifteen tons of food from cradle to grave. At any moment, he or she weighs less than half a per cent of the total stuffed in and varies, most of the time, by just a small amount from that point. I have, in the past half-century, gained three kilograms over my skeletal student self, the weight of half a dozen pound coins a year (and I still weigh less than all but one in twenty of my fellow citizens of similar height and age). Appetite has done its job. The global obesity epidemic comes from the fat controller slipping out of balance by just a tiny fraction.

Appetite is a sixth sense. The balance of hunger and satiety involves dozens of messengers that pass on information about how full the stomach is, how thick the blood has become, how much sugar it contains and how much fat has been laid down. Although many of the body's responses to the new world of excess are implanted in the mind by custom and by habit, genes are also involved in the decisions as to when to lay down the knife and fork.

In 1903 two scientists from University College London found that when food enters a dog's small intestine the pancreas begins to pour out juices that help in digestion. The assumption at the time was that the brain and gut spoke to each other through the nervous system. With an anaesthetised animal, they cut the nerves involved – but the response stayed the same. Later experiments in which the circulations of two subjects, one hungry and the other satiated, were linked showed that a chemical message – a hormone – was carried by the blood.

Such things are the products of genes. Some hormones say

when enough is enough, others when it is not. Each can be struck dumb by mutation and may as a result lead to gluttony and overweight. Such errors are rare, but as hundreds are known they may be behind many cases of severe obesity. The double helix also leads to variation in the lust for food among the general population.

Even in their first few months, identical twins are similar in how hard they suck and how much they enjoy their time at the breast. Now we know some of the genes involved. The FTO ('fused toes') gene in mice causes the digits of animals with a single copy to fuse together (those with two copies die before birth). The version found in ourselves tells us instead when to push the plate away. The intensity of its message is determined by a tiny change in the DNA. Around half of all Europeans bear a single or a double copy of the letter A at a certain point, and the remainder the alternative letter T. Adults with two doses of A weigh, on the average, three kilograms more than those with two copies of T, while those with one copy of each are intermediate. For thirty or so other genes certain variants are more frequent in the fat than in the thin. People who drew an unlucky hand for many of them weigh up to seven kilograms more than those who inherit the low-appetite alternatives. Such differences act at once. Babies with two copies of the risk variant of FTO weigh the same as others when they are born, but within a few months put on an extra quarter of a kilogram.

Greed is the spur but some people have sharper spurs than others. For reasons of constitution and of habit, they find sweet, creamy and fatty foods – mother's milk in solid form –

irresistible. To call them addicts is too loaded a term, but a real tie exists between the effects of such items and those of certain drugs (as anybody who has experienced marijuana-related 'munchies' will know). Even a cigarette makes some people feel peckish. The molecular circuits in the brain that react to narcotics overlap with those that light up in response to sweet and fatty food. Obsessive over-eaters hence suffer some of the sensations of those in thrall to more sinister chemicals.

Genes and hormones explain why some people are fat and others are thin but they are no more than part of the story. Experience, opportunity and habit are also much involved. Satiety, like sanctity, can as a result be conjured up with some simple tricks.

People told to imagine that they have eaten thirty chunks of cheese, or thirty sweets, eat less Cheddar or less chocolate when offered the real thing on a plate. Consumers of imagined cheese eat normal amounts of confectionery and vice-versa. In much the same way, a three-year-old given a normal or a heaped plate of spaghetti will take the same amount from each. Two years later, the same infant will eat more when given the larger portion. Adults are just as gullible. Students fed bowls of soup, supplemented with extra pumped in through a tube in the bottom of the dish, swallow far more than normal. They had learned long before not to stop until their plate was empty, and if it magically stayed full they continued to slurp. Fast-food chains have been quick to take advantage. In the past forty years the size of American hamburgers has increased by up to five times and its citizens have risen like heroes to the challenge. Modern cookbooks, too have, increased the weight of the recommended ingredients compared to those published half a century ago. Even the sizes of the portions in the Last Supper are two thirds larger in the artworks of today than in those of the tenth century.

Brillat-Savarin was convinced that a good meal acted directly on the brain: as he said, 'Tell me what you eat and I will tell you who you are.' For him – as for many others – food brought contentment. At the other end of the emotional scale, for the unhappy, even salt loses its savour. Misery or hopelessness may dampen appetite almost to extinction. So strong is the effect that certain anti-depressants work in part because they sharpen the sense of taste.

One disorder leads to a complete loss of appetite and to serious problems, in young girls most of all. Anorexia nervosa, as it is called, is accompanied by great mental and physical changes. In its first stages there may be a sense of elation, or even of communion with God, but that may be followed by despair. The condition kills a higher proportion of its victims than does than any other psychiatric illness.

The priesthood has long been aware of the power of hunger. Moses twice fasted for forty days and forty nights, first before he was given the Tablets of the Covenant and again when, after his return, he found the Israelites prostrate before a golden calf. King David did the same when his son was at death's door, and later episodes were called for to avoid defeat in battle, to celebrate victory or to divert the judgement of God. John the Baptist ate locusts and wild honey, and Paul starved on the road to Damascus. Jesus himself abstained in the desert and among the temptations placed before him was an offer to transform stones into bread. Such rituals are reflected in the fasts of the Jewish Day of Atonement and in the Christian Lent.

In the sixteenth century the habit of starvation as a sign of piety reached absurd levels, among young women most of all. Anorexia mirabilis, as it was called, led to aberrant behaviour. Catherine of Siena ate no more than a sanctified wafer and a spoonful of herbs each day and if forced to take more would use a twig to extract it from her throat. Now and again she varied her diet with pus from the sores of the afflicted. Wilgefortis, the female saint who with divine help grew a beard to avoid marriage, may have been in the last stages of starvation, for facial hair appears as the body fades away and the balance of hormones becomes disturbed. A few women today still proclaim themselves to be 'Starving for Christ' but not many go so far as to grow beards.

Starvation, voluntary or otherwise, causes the body to call up its reserves. Soon its very substance begins to break down. Many saints died in the 'odour of sanctity', a sweet smell supposed to mark the departure of the soul. The scent is that of acetone, a chemical made in the liver as its capital runs out. In normal times, the substance is a short-term fuel for a variety of organs, most of all the brain. After a few hours of hunger the body begins to summon up reserves from the liver. They last for no more than a day or so and when they have been used up further stores from the muscles are pillaged and fat is burned to provide energy. Once even these reserves are exhausted, demand moves to the muscles themselves, whose protein is broken down. When this last source of glucose is gone, hallucinations set in, the heart and kidney fail, and death, with or without a divine aroma, ensues. Healthy people take about two months to die. Some last longer, but claims of extended fasts are without exception fraudulent. Catherine of Siena's diet of wafers and herbs was no doubt a fantasy and today's Breatharians, who claim to survive with no nutrition at all because they pick up energy from the sun, are not taken seriously even by food-faddists.

At the time of the Reformation certain churchmen began to look askance at the sacred anorexics, together with those who walked barefoot, or went in for flagellation or solitary confinement. Such actions smacked of salvation by physical, rather than by moral, means. The reformers appealed to the words of Timothy that criticised such behaviour: 'Now the Spirit speaketh expressly, that in the latter times some shall depart from the faith, giving heed to seducing spirits ... Forbidding to marry, and commanding to abstain from meats, which God hath created to be received with thanksgiving of them which believe and know the truth.'

The priestly rules of diet were, they decided, against God's wishes. It also rankled that while the laity had to go hungry on fast days, their clerics indulged in banquets in which they fed on fish moulded to look like meat, or fake eggs made of salmon roe and almond milk. Some even interpreted the laws to mean that they could have full meals as long as they ate them outside the refectory.

The new movement was opposed to any attempt to gain purity by physical self-denial, be it by starvation, by solitude, or by lack of shoes. Instead its founders preached 'justification by faith', the notion that the hunger for holiness should be in the mind rather than the body. All attempts to mortify the flesh were, of their nature, sacrilegious.

An act of defiance against the dietary rules was the first course of what was to become a revolution. The Swiss reformer Ulrich Zwingli was certain that the Church needed purification. In Zurich, in 1522, on Ash Wednesday – the start of Lent and a day set aside for self-denial – he persuaded a dozen of his followers ostentatiously to cut up and eat two large smoked sausages. That insult attracted the ire of the local bishop (not lessened when the turbulent priest then demanded, at the request of his own mistress, that the clergy be allowed to marry). In his sermon 'On the Choice and Freedom of Foods' Zwingli argued that Christians were not bound by ecclesiastical rules about what, when and how much they should eat. St Paul himself had written that 'Meat commendeth us not to God: for neither, if we eat, are we the better; neither, if we eat not, are we the worse.' Instead the faithful should have confidence in Christ alone and should partake of a simple diet at all times. The authorities were outraged and became more so when he insisted that the Mass was a meal and not a miraculous transformation of bread into flesh and wine into blood. In his church the congregation sat not in front of a gorgeous altar, but at tables set with wooden cups and plates and laden with a meagre repast that emphasised its literal, rather than symbolic, nature. Soon, Zurich gave up feasts and fasts altogether, shrouded off its saints, and transformed the city's monasteries and nunneries into homes for the sick and the indigent.

All this was not enough for the real purists. They demanded

an even more literal interpretation of the Good Book. The Anabaptists, for example, insisted that baptism should be delayed until adulthood, when the candidate understood what the ceremony signified. They also disagreed with Zwingli about the rules of diet, for they insisted that the communion wafer should be eaten from the fingers of those who receive it rather than being inserted into the supplicant's mouth by the pastor. Infuriated by such heresies, Zwingli issued an edict against such practices and in 1527 the first Anabaptist martyr was bound and thrown into a river, where he drowned in his 'third baptism'.

As disagreement spread, the reformist sects entered into more and deadlier quarrels. War followed war, and in the complicated series of alliances that followed the army of Zurich blockaded their enemies to starve them into submission. Then it was the turn for the Catholics to attack and in 1531 Zwingli was killed in a battle on the outskirts of his native city.

His was but one of the many deaths that have accompanied the ancient disputes about the laws of diet. Those imposed on the Israelites by Leviticus were abandoned by the early Christians on the grounds that theirs was a system of principle rather than practice. The Reformation returned to that detestation of unnecessary ritual. Now, the bloody battles to establish the true significance of bread and wine have been replaced by an acceptance that while many of the rules of the table are no more than traditions others are coded deep into our very frames. Thanks to molecular gastronomy our dietary habits have changed even faster than in Reformation times.

What was once unacceptable is now commonplace and excess rather than deficiency has become the norm. Science can say little about food as fuel for the soul rather than the body but it has at least begun to understand our earthly appetites. That may one day help solve the problems that have arisen as famine disguised as feast has spread across the world.

CHAPTER NINE

THE COLOUR OF THE CRYSTAL

William Blake, *The Vision of Eliphaz*

Then thou scarest me with dreams, and terrifiest me through visions.

JOB 7:14

'And every one had four faces, and every one had four wings. And their feet were straight feet; and the sole of their feet was like the sole of a calf's foot: and they sparkled like the colour of burnished brass ... they four had the face of a man, and the face of a lion, on the right side: and they four had the face of

341

an ox on the left side; they four also had the face of an eagle ... behold one wheel upon the earth by the living creatures, with his four faces ... their appearance and their work was as it were a wheel in the middle of a wheel.

'As for their rings, they were so high that they were dreadful; and their rings were full of eyes round about them four. And when the living creatures went, the wheels went by them: and when the living creatures were lifted up from the earth, the wheels were lifted up. Whithersoever the spirit was to go, they went, thither was their spirit to go; and the wheels were lifted up over against them: for the spirit of the living creature was in the wheels ... And the likeness of the firmament upon the heads of the living creature was as the colour of the terrible crystal, stretched forth over their heads above.'

Thus the vision of Ezekiel and thus many other nightmarish, hallucinogenic, disordered – and even glorious – fantasies from then until now (and, with some chemical assistance, I have had a few myself).

In 1962 a graduate student in theology at Harvard Divinity School set out to find the roots of such experiences. In the Marsh Chapel of Boston University, just before the Good Friday service, a group of his fellow students was divided into two. Half drank a healthy shot of vitamin B3 but the rest swallowed an equivalent dose of psilocybin, the drug found in magic mushrooms and a substance synthesised not long before in the laboratory of the Swiss chemist Albert Hofmann (who, many years earlier, had made the first LSD).

The Marsh Chapel event changed lives. Many of those who had taken the drug said that their moral insights had been transformed. Almost all felt a new sense of unity, of transcendence and of sacredness – each an attribute associated with the deepest consolations of prayer.

Three decades later when they recalled that magical moment most of the by then middle-aged disciples felt that it had exerted a permanent influence upon their faith. Some said that for the first time they had understood the true power of the Lord. Although they were no longer legally able to do so, many wished to have a further bite at the phantasmagorical cherry; to use narcotics once more to buttress their convictions.

In 2006 the exercise was, with police approval, repeated with a larger group of pious Americans. The subjects were presented with three hundred queries about their responses, taken from the Pahnke-Richards Mystical Experience Questionnaire. They were asked about ineffability and paradoxicality, and whether they had experienced sensations of oceanic boundlessness and noetic quality. Some saw visions. Two out of three had what they described as a complete mystical immersion and rated the day as among the top five moments of their lives, equivalent, for example, to the birth of a child. Over the next few months their families reported that the participants had become better people. A year later some recalled a 'conversation with God (golden streams of light) assuring me that everything on this plane is perfect' or that 'horror is only an illusion and God lies beneath it all'.

Their sensations were close to those reported by those who spend their lives in contemplation of the divine. Might they share a source?

Devotees insist that when they put their trust in a higher power they ascend into a universe of thought denied to sceptics. They may, in their own eyes, be right; but similar sensations can emerge from simple physical changes in body and brain. Cynics point at the parallels between the actions of saints and visionaries and those of people with schizophrenia, depression and other mental disorders. The philosopher William James, himself a committed Christian, accepted in his 1902 book *The Varieties of Religious Experience* that 'religious geniuses have often shown symptoms of nervous instability'. He spoke of George Fox, the founder of Quakerism, who one day in icy weather near Lichfield heard the voice of the Lord. It told him to take off his shoes and to parade through the town, shouting, 'Woe to the bloody city of Lichfield!' Fox wrote that: 'As I went thus crying through the streets, there seemed to me to be a channel of blood running down the streets, and the market-place appeared like a pool of blood.' He persuaded himself that this revelation arose from the martyrdom of a thousand Christians there in Roman times.

George Fox, thought William James, was an admirable person but a 'psychopath of the deepest dye'. Even so, the philosopher dismissed all attempts to understand mystical insights with an appeal to pathology. 'Medical materialism', as he called it, was trivial: it 'finishes up Saint Paul by calling his vision on the road to Damascus a discharging lesion of the occipital cortex, he being an epileptic. It snuffs out Saint Teresa as an hysteric, Saint Francis of Assisi as an hereditary degenerate.' Biology was not the right tool with which to explore devotion.

James' dismissal of a physical basis for spiritual experience was odd for he had himself experimented with the effects of chemistry on the mental universe. Nitrous oxide had been discovered a century earlier and was valued as an anaesthetic but as 'laughing gas' it was also indulged in as a source of simple fun by Coleridge, by Southey, and many others. In 1874 William James read an obscure pamphlet entitled *The Anaesthetic Revelation and the Gist of Philosophy* which argued that the secrets of the universe were to be found in hallucinogens. He went on to sample the gas and found that it had a dramatic effect: 'The keynote of the experience is the tremendously exciting sense of an intense metaphysical illumination. Truth lies open to the view in depth beneath depth of almost blinding evidence.' Like the students in the March Chapel, he wrote of the event as the strongest emotion he had ever had.

His notes on his mental journey were perhaps a little incoherent: 'What's mistake but a kind of take? ...What's nausea but a kind of –ausea? ... Emotion-motion!!! ... It escapes, it escapes! But – What escapes, WHAT escapes?' Such insights allowed William James to deduce that our normal, rational, consciousness is separated by the filmiest of screens from an entirely different set of experiences. For James the gas was the ticket for a voyage between the real and the supernatural. His brush with chemistry hinted at the deepest foundations of devotion. To him, 'An hallucination is a strictly sensational form of consciousness, as good and true a sensation as if there were a real object there. The object happens not to be there, that is all.'

Phantasms and the divine are old friends. The *Catholic Encyclopaedia* sorts such experiences into several classes. They include corporeal apparitions of the Blessed Virgin and the like, together with imaginative visions, 'The sensible representation of an object by an act of imagination alone, without the aid of the visual organ and with or without supernatural help, as long as they are accompanied by the graces of sincere sanctity.' Some among them must have physical reality, for they are attested by burn marks left by the passage of the devil.

However conclusive such evidence might be, the *Encyclopaedia* accepts that some visions do arise from self-delusion, or from mental illnesses such as melancholia, epilepsy, paranoia or hysterical psychosis. Psychiatrists agree. The *Diagnosis and Statistical Manual of the American Psychiatric Association* attempts to codify and to clarify the confusion that surrounds mental disorder. Its categories include 'Religious or spiritual problems' in an acceptance of at least a partial overlap between the world of dogma and that of medicine.

The Bible is full of such experiences. Daniel, who 'had understanding in all visions and dreams' reminded King Nebuchadnezzar of his nightmare of an image with feet of clay that prophesied the fall of Babylon. The Book of Isaiah explains that its verses are an account of 'The vision of Isaiah the son of Amoz, which he saw concerning Judah and Jerusalem' while Joel speaks of a wonderful time when 'your old men shall dream dreams and your young men shall see visions'. Those creative moments pale when compared to the last book of the Bible, in which John of Patmos' revelations of the end-time reveal a set of numerical obsessions – seven lamp-stands, seven stars, seven churches, seven seals, seven trumpets, seven bowls – and of fabulous monsters (some with seven heads). To match that there are two hundred million horsemen on steeds with heads like lions, and rivers of blood. The book's contents would be at home in the most fevered imagination and George Bernard Shaw compared them to 'the peculiar record of the visions of a drug addict'. The sleep of reason still attracts the gullible to places such as Medjugorje in Bosnia-Herzegovina, where in 1981 six children began to have conversations with the Virgin Mary. Now, thousands of pilgrims arrive each week to see her manifested in the Dancing of the Sun or to visit the Bedroom of Apparitions.

Science, in its banal fashion, makes it possible to study the mind in ways impossible in the days of William James. The visions of saints, sinners, dreamers, drug users or anyone else can now be explored with technology. To do so may not give much insight into piety itself, but hints that at least some of its phenomena are side-effects of the machinery of the nervous system. People are free to follow the voices in their heads but they should realise that, as far as biology is qualified to tell, many have a material origin.

Some scientists go further. They suggest that the entire structure of conviction, from the visitations of spirits to the hope of the afterlife, and from the architecture of the Vatican to the music of Bach, emerges from artefacts of perception. The divine mysteries are no more than a sort of phosphorescence of the brain; an incidental of its normal (or abnormal) function. That organ – like all others – cannot always cope with what existence throws at it. Its mistakes and failures manifest themselves in many ways. Some are interpreted as illusions, others as psychiatric disorders, but yet others as messages from a higher plane.

Many people – atheists and believers alike – are indignant at the claim that their convictions, be they of free will or of divine intervention, are no more than quirks of the nervous system. It is almost impossible to prove or disprove that idea. Even so, hallucination and even mental disorder do play a part in the histories of all creeds and to understand them might hint at what some of the varieties of religious experience may actually be. Science shows that many of our perceptions of reality lie between the ears as much as in the world outside or, for that matter, beyond.

False impressions are part of human nature. Most stockbrokers are convinced that they can beat the market, but over the long term the rise and fall of the index has a random pattern. Wars are often sparked off because politicians on each side are sure that they will prevail, only for one party to be proved wrong. Many of us are certain that we are cleverer, more intelligent and more attractive than our peers judge us to be. A simple game shows how. When offered cash for each correct answer to a series of quiz questions, or the chance of a competition with three others and, under a 'winner take all' rule to rake in four times as much if they win (but nothing if they lose), three out of four men choose the contest, for each is boosted by the sensation that he is better than average (which is by definition impossible). Women have more sense.

As conjurors know, the brain is always ready to fool itself, often in remarkable ways. Someone who has lost a hand may be tormented by a 'phantom limb'. Their nervous system refuses to accept that the structure has gone and interprets it as present, but stuck in a cramped position. Non-existent fingers dig into an imagined palm and patients face intense discomfort as they try, and fail, to unclench their fictional fist. Their distress is genuine. It shows that pain, like many other messages from the external world, is not always a true statement of reality. A simple optical trick – a vision of the unreal – works miracles. The patient hides his stump (which might be on the left) behind a mirror set at right angles in front of him. He looks into it and shifts the image of his clenched right hand until it appears to replace its absent partner. Then he commands both fists to open. His real hand obeys at once. The image in the mirror appears to do the same and the pain from the absent hand goes away.

Sceptics often use such observations to mock what they see as the delusion that human actions are subject to the influence of a higher power. They boast that they at least act in a manner determined by their own free will. Their decisions, be they to make a cup of tea or to blaspheme against the Holy Ghost are made, proudly, of their own volition. Such people deny the existence of an unconscious world. Brain science should give them, as much as their opponents, pause for thought.

From Aristotle onwards, philosophers have been interested in our apparent ability to decide on an action without reference to an external constraint. Theologians, too, speculate about the conflict between free will and a higher power. For Christians free acceptance of the teachings of Jesus Christ is essential to salvation. To neuroscientists the whole issue looks more and more like an illusion. Nobody chooses to breathe in and out, and nobody, sovereign as they might feel, can hold their breath for fifteen minutes. The nervous system overrules their wishes and they are forced to gasp for breath. In fact, all actions, apparently voluntary or not, are preceded by brain activity outside the perception of those who make them. It then becomes almost impossible to separate the conscious from the unconscious mind or to disconnect the inner senses from the apparent intervention of an external agent.

Students wired up to a scanner were asked to sit in front of a clock and, whenever they felt like it, to flick up their hands and to note the exact time when they decided to do so. Easy enough, but about a second before they recorded each decision, a section of their grey matter had already burst into activity; the resolution to act had been made before the actor was aware of it. Some responses are preceded by as much as five seconds of animation, so that well before the subjects feel that they have made a choice the brain has already decided what to do. People instructed to empty their minds as far as possible reduce the activity of the higher centres by no more than about a tenth below the level achieved while they are working out complex sums. The brain, it seems, never stops thinking however empty of thought it might seem.

Such blurring of the boundaries between conscious and unconscious means that some people come to feel that they are under the direction not of their own free will but of a force from outside.

The first hint of the overlap of the two came from the

eighteenth-century Swiss naturalist Charles Bonnet, better known for his discovery of virgin birth in aphids. His grandfather, almost blind but mentally lucid, complained that he often saw faces, birds, tapestries, and patterns and had no idea from whence they came. Bonnet realised that they were the products of the brain itself, merged with signals from the real world. Sometimes, his grandparent saw a deceased friend at rest in a real armchair; a common delusion among the bereaved, almost half of whom report the sense of a presence of the departed. His aged nervous system, deprived of its normal stream of information, generated images of its own from the store it had tucked away over the years and mingled them with others from outside.

In 1894, the Society for Psychical Research carried out a Census of Hallucinations. Seventeen thousand Britons were asked whether they had ever had, when awake, a delusion of voice or vision. Two thousand replied in the affirmative. A Mr Eggie wrote that: 'On October 5th, 1863, I awoke at 5 a.m. ... I heard distinctly the well-known and characteristic voice of a dear friend, repeating the words of a well-known hymn. I have always thought it remarkable that at the very same time, almost to the minute, my friend was seized suddenly with a mortal illness.' Others were more sceptical. From Mr W.S.: 'I became painfully anxious, and sat down for a minute ... and then I saw, floating, as it were, between me and the mantel-piece, the upper half of my father's body ... I looked steadily at my half-ghost, and saw how a spot in the mantel-piece, a knot in the wainscot, &c., &c., had combined to produce the spectral appearance.'

William James was impressed by the Census (but less so by the claim that 'between deaths and apparitions of the dying person a connection exists which is not due to chance alone') and the research was well ahead of its time. Our isle is still full of phantoms. Millions of Britons report that dead relatives have spoken to them or that they have heard – like Adam and Eve – the voice of God. An even higher proportion of Americans is so persuaded (as were Freud, Napoleon, Churchill, Hitler, Gandhi, Shelley and Keats). Sometimes the supposed sights or sounds are side-effects of our normal functions (as when people mutter to themselves) while others have causes as prosaic as radio transmissions picked up by dental fillings.

Self-deception can involve the body itself. The opera director, doctor, and fervent anti-monarchist Jonathan Miller tells of an occasion when, looking down his large and republican nose at an enthusiastic crowd as the Queen drove by, he found that his own hand had decided to wave frantically to the Royal Personage, against orders from Central Command to stop. This 'anarchic hand syndrome' is reflected in *Dr Strangelove* and for some unfortunates the errant appendage steals food from a neighbour's plate or even tries to strangle its owner.

Certain sections of the brain predict the movements we are about to make while others interpret where our arms, legs and other body parts are placed at any moment. Most of the time, the former take precedence and we take no conscious notice of the latter. Sometimes, though, the streams become confused, to give the troublesome feeling that active movements of an arm or leg are a response to an external, rather than an internal, agent. In extreme cases people imagine that they have an extra limb, or even that their whole body has been duplicated to generate a 'doppelgänger'. Those who have seen the latter include Abraham Lincoln, who was convinced that he had glimpsed a double image of himself in a glass, and Goethe, who was approached by his twin perched on a horse. The duplicates are sometimes identified as harbingers of death that persuade the body to enter a foreign realm. As the sensation fades, some report that they themselves have come back from beyond the grave.

Many of the pious do not wait for the divine afflatus to make its presence felt of its own accord but, like William James, use a short-cut. Deep mental absorption can persuade the brain that an inner thought has an outer source. Priests of many religions spend solitary hours in darkness or in silence. Such experiences may activate the pineal gland at the base of the brain. Descartes believed that to be the seat of the soul. Be that as it may, the structure is the source of melatonin, a chemical concerned with sleep and wakefulness. Those who meditate may have more of it than others, with a parallel shift in mental condition.

Such biochemical changes are sometimes inflicted upon people who have been forced to meditate against their will. Solitary confinement is among the several forms of torture to which the United States has become addicted in recent times. In the nineteenth century the punishment was seen by the Quakers of Philadelphia as a social advance and its use was designed to allow prisoners to 'reflect on their bad ways, repent, and then reform'. The idea spread to Europe. Soon it became clear that far from acting as an agent of reflection and reform, separation from the world caused severe mental upset. Charles Darwin, after a visit to such a place, wrote that the inmates were 'dead to everything but torturing anxieties and horrible despair'. In time, the practice more or less died out.

Two decades ago it began to be used once more in America's so-called 'super-max' prisons. More than fifty thousand of the nation's citizens are now held in isolation in tiny cells with no windows and constant dim light, fed through a slot in the door. They are also kept in complete silence. Some have been locked in for two decades. The effects can be dire. Seven out of every ten prison suicides are among those kept in solitary and within a few days many of those thrown into the cells begin to see faces, to hear voices and to become fixated on the idea that they are under the control of an unknown force.

After reports of brainwashing by the Chinese at the time of the Korean War, psychologists began to study the effects of isolation. They soon found that even a few hours could spark off apparitions. Among students placed in pitch blackness and complete silence the susceptible saw faces or felt an 'evil presence' within as little as fifteen minutes. With such extreme lack of input, the slightest internal perception – a heartbeat, a muscle twitch, or a heavy breath – is interpreted as a message from elsewhere; perhaps, indeed, from the gods.

The brain can also be assaulted directly. The deep breaths of Eastern mystics (and of church choristers) purge the blood of carbon dioxide, make it more alkaline, and fool the nervous system that oxygen is available even as the vital gas runs short. Some of the techniques used are arduous. I once visited an encampment of German Hobby Indians in which young men, dedicated followers of what they see as the pure life of Native Americans, danced in grim silence for hours until they collapsed into a trance. As the brain gasps for oxygen, the limbs tingle, the head spins and, now and again, the mystic finds a moment of ecstasy.

That state can be reached with less effort through chemistry. Narcotics and the supernatural have long gone together. An eight-thousand-year-old cave painting from Algeria shows a priest about to eat a mushroom, the Scythians used cannabis to make contact with the hereafter, and 'ambrosia' – the sacred drink of the pre-Hellenic tribes of Greece – was, Robert Graves suggested, a distillation of psilocybin-filled fungi. Siberians were said to drink reindeer urine to sample the same stuff, for it became concentrated there because of the animals' taste for fly agaric. Even in the Bible Lands, chalices found at Philistine sites include remains of a hallucinogen, perhaps nutmeg imported from India.

The divinity students in the Good Friday Experiment saw their experience with psilocybin as a turning point but William James found the stuff less inspiring: 'I took a small dose at 6:30 a.m. and had nothing but nausea & diarrhoea till 4 the following a.m., when I remember I vomited for the last time.' My own limited exposure to the peyote cactus, a source of mescaline (which is related to the chemical used in the Marsh Chapel) also involved nothing more meaningful than some colourful illusions, a slight queasiness, and the sense that the whole event had gone on for too long. That prosaic response was not improved by the babble of Colorado hippies in our geodesic dome who were having a better time on highways of the supernatural than was I. There was no oceanic boundlessness, no sign of sanguinary streets and I would not have recognised noetic quality if it had blundered into the room. Inborn cynicism, a biological inability to bow to the drug, or a simple failure to enter a cultural milieu alien to a scientist – even one as callow as my younger self – were all no doubt to blame. Whatever the reason, my reaction was far less dramatic than those of people primed to expect a mystical insight (and I would not be surprised to learn that many of those 1970s hippies have since become born-again Christians).

That decade was an era of altered perception. The gullible often persuaded themselves that narcotics could put them in touch with another world. They had no interest in conventional dogma – Christianity least of all – but found prophets in other places. Twenty years earlier the writer and heroin addict William Burroughs had travelled in South America. He regaled his friend Allen Ginsberg with letters about his drug-fuelled adventures. Their publication as *The Yagé Letters* (yagé was a plant extract that he imagined would give him telepathic powers) led to a fad for psychoactive substances as a key to the infinite. Carlos Castaneda's series of books on the subject sold eight million copies. They began in 1968 with *The Teachings of Don Juan*. His tales of adventures in the other world under the influence of hallucinogens provided by a

Yaqui Indian spirit-guide from Mexico are in fact works of fiction (even if one volume did earn him a PhD from the University of California at Los Angeles). Even so, they had an extraordinary effect on the youth of those days.

Shamans act as bridges between the natural and the supernatural. Their trademark image of the plumed serpent joins the upper world of the birds with the underworld of the snakes. The shapes seen by the intoxicated mystics were much the same wherever they lived, even if they interpreted them as reptiles in some places and jaguars in others. Two thousand-year-old rock art that shows fierce animals fused with humans is found from Mexico to the south-western United States.

Such seers can, some say, heal the sick, predict the future, and journey in the realm of the spirits. Their ecstatic state is brought on by chemicals. In Latin America they include strong tobacco, cacti filled with mescaline, and *ayahuasca*, a mixture of a certain vine and of other plants that releases a powerful hallucinogen (and causes severe nausea, which adds to the profundity of the event). The remains of a ritual drink have also been found in thousand-year-old vessels excavated from the enormous Native American site of Cahokia, in Missouri. It was brewed from a species of holly and contained, in a precursor of modern habits, lots of caffeine – but also substances that lead the drinkers to vomit. Early European explorers noted that such rituals accompanied football games and were also used in spiritual ceremonies.

The names given to the plants involved showed the regard in which they were held. The Aztecs called the cacti *teonanácatl*, or 'God's flesh', while the fungi were referred to as the 'most holy of lords'. To match that, the Spaniards referred to the seeds of the morning glory as the *semillas de Virgen*, the seeds of the Virgin Mary, as a hint that they recognised their mystical powers. The habit lives on in Brazil, where elements of shamanism have been wedded to Catholicism to generate a creed called Santo Daime, which mixes Christianity with *ayahuasca*. Its adherents often see Jesus himself. That hybrid dogma has spread across the world. The Constitution's guarantee of freedom of worship means that its rituals are allowed in the United States although they are illegal in most of Europe.

The drugs involved are potent indeed. Even tobacco in its native form contains so much nicotine that those who smoke it can see only in black, white and yellow and perceive people around them as ghosts. It produces a sense of flight, together with vivid dots, grids, arcs and tunnels, sometimes accompanied by an apparent death followed by a struggle back to life. Magic mushrooms and cacti cause their devotees to interpret trees or houses as having human features (and some people continue to suffer such delusions twenty years after they take the stuff). The brain is always ready to see a face, that of the Man in the Moon and Ezekiel's wheeled creature included, and a mind addled by chemistry may perceive a random aberration as a divine countenance.

Hallucination has moved on, from chemistry to physics. The God Helmet uses a snowmobile crash-hat decorated with electromagnets arranged to be close to particular areas of the brain. Many who try it sense a presence in the room, which some interpret as that of – who else? – the deity. Others experience simple mental elevation (although attempts to awaken Richard Dawkins' internal seraph have not yet succeeded). Cynics claim that the result is obtained even when the magnets are not switched on.

The effects of chemicals upon the nervous system overlap both with the visions of the pious and with mental disorder. The task of disentangling quite why they work and how we respond to the internal and the external worlds will take years and some of the many claims made by those who study the subject are simplistic or even fatuous. Even so, as the biology of the brain is unravelled, some hints have emerged. Its own chemical messengers are intimately involved.

The nerve transmitter serotonin was once thought to be a simple 'happiness molecule'. In fact it acts in a wider fashion to restrain the brain's higher centres' tendency to run to extremes, be they of hope or despair, by filtering the information that comes in from outside. Drugs such as LSD or peyote block its receptors, release the mental brakes and allow chaos to reign. The substance suppresses emotion and memory and reduces the extent to which different parts of the grey matter pass messages to each other. Cannabis contains another compound that mimics a transmitter. It causes nerve cells to remain switched on for longer than they should and to respond when they ought to be at rest. That gives a perception of bright colours, swirling patterns and harmonious music. Heroin, in contrast, mimics a chemical called dopamine that gives a sense of reward and alleviates pain. Its ability to calm and to gratify soon leads to addiction.

Each transmitter family has many members, each performs several tasks and each is associated with a variety of enzymes, of receptors and of channels on the surfaces of cells. Together they change behaviour, interact with drugs, and shift the threshold between the real and the imagined. A step too far over that boundary leads to mental illness.

That was once blamed on demons. Jesus drove out the evil spirits that had taken control of an unfortunate who was 'kept bound with chains and in fetters; and he brake the bands and was driven of the devil into the wilderness'. They entered the Gadarene swine, which ran down a steep place into a lake and were choked. That robust approach lasted almost to the present day. Some sufferers were burned because they had made a Satanic pact while more were exorcised to drive out the Evil One's emissaries.

Medicine has, in its own way, expelled many of the malign spirits, but millions of people still live in a place ruled from within their own disordered skulls. Often, they interpret their delusions as messages from a higher power. As psychiatry tries to control their problems it has, on the way, gained some insight into the overlap between mental illness and mystical experience.

As William James pointed out, attempts to label such sensations as no more than failures of brain function can be simplistic. Even so, some almost beg to be explained in such terms. Epilepsy, schizophrenia, and more have long been used to rationalise the visions of the devout. The notion that a conversation with God involves no more than a short circuit in the nervous system at least allows that perception to be studied by science.

Saul, persecutor of Christians, when on the road to Damascus, saw an intense light and heard a voice that commanded him to preach the Gospel. He was struck blind ('when his eyes were opened, he saw no man; but they led him by the hand, and brought him into Damascus. And he was three days without sight, and neither did eat nor drink'). Muhammad, too, fell prostrate and, as he did, told of voices and extraordinary sights. In more recent days, Joseph Smith, founder of Mormonism, lost consciousness and found himself unable to speak. He noted that 'When I came to ... I found myself lying on my back looking up at heaven.' Some of those descriptions are consistent with an attack of epilepsy. Even the idea of a single god has been ascribed to the condition. The pharaoh Akhenaten reigned in around 1150 BC and raised a minor sun-disc deity, Aten, to the supreme position, perhaps, some speculate, after an epileptic vision of a bright sphere.

The Greeks called epilepsy 'the sacred disease' as it was assumed to be due to divine possession. An electrical storm in the brain causes victims to fall rigid to the ground. They shout, thrash their limbs, or say that they see bursts of lightning. An attack can be sparked off by heat, or by bright light.

Schizophrenia, another grievous illness, is often invoked to explain mystical encounters. It affects around one person in a hundred. Members of racial minorities (whoever they might be) and city-dwellers are each more at risk, perhaps because the solitary and the disturbed gravitate to places where they can sink into an anonymous mass. Many prophets have been strangers in strange lands and have found their way to great cities to preach.

For most people, dreams are less convincing than is reality, but for schizophrenics the opposite is true. They are under siege from the supernatural. Problems often begin in the late teens, with anxiety, depression, bizarre fantasies of persecution by an external agent, or of the sound of urgent voices. Thoughts that seem innocuous to most people – about sex, or even about a decent meal – turn into torment. Many victims attempt suicide, and some succeed. No more than one in six recovers within five years. Few can hold down a job, many are homeless and people with the condition die twenty years younger than average.

Often, the symptoms reflect a wish to purify a polluted life. Obsessions may play a part. The stereotyped nature of certain church ceremonies are not much different from the actions of those who make, like many schizophrenics (and, for that matter, John of Patmos), repeated use of a set of colours, numbers or actions to build a place that feels safe to them even if their activities have no results visible to anyone else.

Devout patients are at higher risk of succumbing to such behaviour, often because of a sense of guilt that they are unable to live up to the stringent commands of their deity and that punishment, now or in the afterlife, cannot be far away. Some think that they are possessed by the Devil or have committed unpardonable sins. They may obey biblical injunctions to pluck out an eye or to castrate themselves, or refuse to take drugs because that seems to break their tie with God. Many insist on their own divinity or imagine that if they commit suicide they will be healed in heaven. The United States has five times as many people with a morbid sense of saintliness as does our own secular land. As, a century ago, the Anglican Church began to lose its power, the thoughts of British schizophrenics began to concentrate less on the Bible and more on sex.

The illness was once blamed on rejection by a mother, but that was a fantasy. For some people inheritance is involved. A great variety of genes has been implicated, many said to be damaged by the insertion or loss of short sequences of DNA, but in truth there has been almost no success in tracking down the actual errors responsible. Some of the supposed candidates are active not just in the brain but in the immune system. Perhaps they interfere with the brain connections made as the body shifts into adulthood but even that is not certain. About the biology of schizophrenia, as for so many mental disorders, we know almost nothing.

Even so, its symptoms are a powerful reminder of the overlap between mental disturbance, the effects of drugs, and intense devotion. Amphetamines and cocaine produce sensations that may resemble those of the illness and those who have the condition are more susceptible to such substances. Cannabis users face twice the average risk of schizophrenia, both because the stuff induces visions and because some people with the condition find that it makes their lives easier to bear. Almost every schizophrenic smokes, for they are medicating themselves with nicotine, which interferes with the dopamine system. PTP or 'angel dust', together with ketamine (an anaesthetic used by clubbers who appreciate its dreamlike state whatever the damage to their nervous system) and nitrous oxide, William James' ticket to the supernatural, have also been used in treatment.

Dangerous as the illness can be, some suggest that in its mild form it gave us our insights into the world around us; that 'great wits are to madness near inclined'. Einstein's schizophrenic son and James Joyce's similarly affected daughter are often cited, as is the claim that around a third of the children of mothers with the disease who are adopted into healthy households develop an intense interest in religion.

Whatever the truth of such notions, many saints and seers have been said to show its symptoms. Teresa of Avila, founder of the Discalced Carmelites, had repeated ecstasies, raptures and visions and, in her youth, a four-day episode in which she was struck rigid (a symptom of a certain form of schizophrenia). So intense was her state that she was assumed to be dead and her grave was dug. Even so, the idea that large numbers of holy figures suffered from it is now less accepted than once it was. Visionaries like St Joan, who began to hear voices when she was thirteen, do not feel the despair shared by many patients. Instead they find themselves in a trance-like conversation with the divine. Even in simple societies, people with the condition are seen as ill rather than possessed of supernatural powers, while shamans – aberrant as their acts may be – attain their state with the help of narcotics and are for much of the time normal.

Perhaps, in the days before science, those who had hallucinations, whatever their cause, could provide an explanation for the world that impressed their fellow citizens, who saw their statements as messages from the gods. The remarkable parallels between the supernatural experiences of members of cultures that have never been in contact hint that they may share a source in the human mind.

Now and again even I, as a prosaic scientist, have visions; not of mysterious empires above the skies, or of a return from the dead, but of an imagined cosmos inside my head. Today's response is to look for physical explanations but in earlier times enlightenment was sought in another world.

The twelfth-century mystic and composer Hildegard of Bingen had the same problem: 'I saw a great star most splendid and beautiful, and with it an exceeding multitude of falling stars ... and suddenly they were all annihilated, being turned to black coals ... and cast into the abyss.' Manuscripts that illustrate her revelations show saw-toothed lines that look like the sunlit parapets of a defensive tower. My symptoms are trivial in comparison, but she and I had the same condition.

My own attacks of migraine are restricted to a brief 'scotoma', a bright but painless circular image that pulses and spins and can wipe out part of the visual field. Around one person in six sees that illusion at some time. The condition was first described by the Cappadocian physician Aretaeus two thousand years ago. The bright lines and lights he interpreted as rainbows, while others see stars, or flashes, or mosaics that undulate like a magic carpet. Visions of battlements, a zigzag line that blocks off some of the field of view, are common. A few people interpret their patterns as herds of animals, or crowds of people. For many, the pulse warns of an imminent headache, of a bout of vomiting, or of a false sense of touch, sound, taste or smell but from these I am spared.

Others imagine that they are falling, that their necks have stretched, or that their arms or legs have ballooned or shrunk. 'Lewis Carroll syndrome' is named after the great author, who without doubt suffered from migraines. In his diary he notes that he had experienced 'that curious optical effect of seeing fortifications'. A sketch in his notebook shows a man with part of his head and his upheld hand missing, just as expected from someone whose visual field has been interrupted by a scotoma. That strange failure of sight might even have given Carroll the idea for the Cheshire Cat, the creature that disappeared until just its smile was left. The size aberration provided the notion of an Alice who could shrink or grow while her miraculously elongated neck was part of the same set of symptoms. Her fall down the rabbit hole into Wonderland may itself have been a reflection of the tumbling sensation reported by some of those with the syndrome.

A few people blame dietary triggers such as chocolate, cheese, or (as in Wonderland) mushrooms for their attacks while others think that damp weather or a persistent wind such as the mistral is the cause. Twice as many women as men suffer from migraine and the condition can run in families. As usual, a variety of genes has been implicated, and – as usual – there is little evidence that any of them are important.

The symptoms emerge from a failure to process sensory information, be it pain, taste, smell, sound, light or the position of the limbs. In an unexpected echo of the lessons of Don Juan, the hallucinogen ergotamine, found in a mould that grows on damp grain and was once blamed for demonic possession, is used to treat severe cases.

Some migraineurs speak of an aura: a magical, an almost divine, sense of calm and peace just before an attack. Dostoevsky wrote of his own episodes that 'There are moments, and it is only a matter of five or six seconds, when you feel the presence of eternal harmony ... During these five seconds I live a whole human existence, and for that I would give my whole life and not think that I was paying too dearly.' Hildegard of Bingen felt much the same. Like Lewis Carroll, she interpreted her visions as battlements; in her case those around the City of God. She was convinced that they were of divine origin.

The Tukano people of South America speak of similar experiences, but they take the bright circle as it spins to be a message from the Sun-Father, while the San of southern Africa interpret the hallucination as a rope to be climbed to the sky. The ramparts sketched in Hildegard's illuminated manuscript resemble certain shapes in African rock art from ten thousand years ago. Such phantoms may even reside within the zigzag lines scored onto blocks of red ochre a hundred thousand years before that.

One perception common to migraineurs, to those with mental disorders, to drug-users, to shamans and to mystics – and to Alice in her rabbit-hole – is of a sudden emergence from a gloomy passage into brightness. In Genesis, one of God's first acts is to divide the light from darkness. In the Book of Job the divinity 'discovereth deep things out of darkness, and bringeth out to light the shadow of death'. Isaiah foretold the birth of the Saviour when he said that 'The people that walked in darkness have seen a great light: they that dwell in the land of the shadow of death, upon them hath the light shined.' Much later, Jesus himself stated that: 'I am the light of the world: he that followeth me shall not walk in darkness, but shall have the light of life.'

Eight million Americans have seen the divine light, and many refer to the experience as a rebirth. Some feel that they have seen the radiant gateway to heaven. The idea of a gloomy and hidden place from which the only escape is through the supernatural is said to be in the cave paintings at Lascaux, in the southern African San's conviction that those who die pass through a tunnel into the afterlife, in the Navajo idea that the soul makes its way to heaven through a hollow reed, and in Dante's descent into, and return from, the underworld.

In the pursuit of the glorious life to come the priesthood has, again and again, given that transition a physical reality. The Mayas spoke of an underworld, a middle-world and an overworld. The sunless universe below could be reached through flooded sinkholes in the Yucatán limestone. They made huge models of such places. Pyramids, with chambers within, were built on top of each other as the centuries rolled on. The bigger they became, the more sacred was the innermost sanctum. Their rulers carried messages between the three worlds and as each was crowned his throne was lifted onto the summit.

Medieval cathedrals, with their shady cloisters matched with a blaze of stained glass, show the same movement from gloom into glorious colour, with the priest elevated above his congregation. Seville Cathedral – once a mosque – was for a time the largest such structure in the world. Those who planned it declared: 'Let there be a church so beautiful and so great that those who see it built will think we were mad.' It was, with its eighty chapels, and its contrasts between pools of shade and patches of sunlight, designed to overwhelm the senses of those who entered and it still succeeds in so doing. With the Reformation, emphasis moved from sight to sound; from lavish edifices to the wonders of counterpoint. Each can still arouse the transcendent even in the most determined sceptic and each, in the end, works its magic through the infinite complexities of brain chemistry.

Much as I appreciate both ecclesiastical architecture and Johann Sebastian Bach, when it comes to their spiritual, rather than their physical, power I have, in spite of the scotomas, a blind spot. Brain science sheds little light on why I am denied an experience so central to the lives of others; and its failure reminds us how little success technology has had in helping us to understand the workings of the inner angel that makes its home within every nervous system.

UNTO OTHERS

William Blake, *Albion before Christ Crucified on the Tree of Knowledge of Good and Evil*

Canst thou by searching find out God?

JOB 11:7

When I first came to the Galton Laboratory at University College London – which was, I realise to my dismay, forty years ago – I was approached by a stooped middle-aged American with a ragged beard and wild hair who asked me with great intensity whether I had a hot line to Jesus. I explained that I possessed no such facility and he moved on. Over the next few months his behaviour became odder and odder and he took to shouting in the corridors about the urgent need to make a connection to the Saviour.

UCL in those years – unlike today, when everyone is tediously sane and spends their time filling up forms – was notable for its eccentrics (many of whom were major scientific figures) and nobody took much notice. George Price – for that was his name – then disappeared from the scene. In January 1975 came the dreadful news that he had cut his throat in a dismal slum a few hundred yards away from where the laboratory then stood.

His despair emerged from a clash between a religious obsession and his research on the evolution of behaviour. Price had studied physical chemistry at the University of Chicago. After a brush with thyroid cancer, a chequered career with IBM, and a failed marriage, in 1967 he moved to Britain. His interests shifted from chemistry to computers and at last to theoretical biology. Soon he became interested in the laws that control animal societies. As he saw, they involve tensions between collaboration and conflict, between immediate satisfaction and delayed reward, between honesty and deception and between the pleasures of a solitary existence as against the urge to be part of a group. Such choices permeate biology, from the evolution of sex to that of ecosystems. They are also the raw material of most faiths.

Soon after his arrival in Britain, Price wrote to the UCL theoretical biologist Cedric Smith. Smith was a convinced Quaker and pacifist (and for many years I occupied what had once been his office). The American's rigorous approach impressed Professor Smith and he was offered an honorary post in the Laboratory. Soon he began to explore how mathematics can help to understand how to respond to another's actions and to determine just when it pays an animal to cooperate with, to defer to, to retaliate against, or to beat a hasty retreat from, other members of its species. To do so it must speculate, consciously or otherwise, about the future. Its decisions depend on the balance of advantage and disadvantage, on the delay before a pay-back and on the chances that it will meet the second individual again. Kinship may be involved, with rewards bestowed upon a relative rather than upon the altruist itself. The need to guard against cheats also plays a part. With a few substitutions of 'minus' for 'plus' George Price found that his equations for mutual aid worked just as well for conflict, vengeance, aggression and spite.

His results are at the roots of much of the modern science of animal behaviour and are used by students of ants with their sterile castes, of birds that help others bring up their chicks, of the vicious struggles between chimpanzee troops and even of the ability of bacteria to clump together when they become abundant enough to allow them to link arms and form a film over teeth or wounds.

Such habits have often been cited by those who hope to explain patterns of human society. The Bible itself is fond of animal archetypes, with its industrious ant, sturdy ass, fierce lion, and so on. Bees attract the liberal and optimistic (Samson's riddle that 'out of the strong came forth sweetness' struck him when he found a swarm in a dead lion); ants, the conservative and choleric (as in Proverbs' injunction to the idle, 'Go to the ant, thou sluggard; consider her ways, and be wise').

To pick anecdotes from animals to justify our own habits is naïve at best; as Alexander Pope put it: 'Thus Ants, who for a Grain employ their Cares,/Think all the Business of the Earth is theirs,/Thus Honey-Combs seem Palaces to Bees,/And Mites imagine all the world a Cheese.' So vast is the diversity of their habits that with a little effort it is possible to find an excuse for almost any human activity. George Price looked instead for general laws that might unite apparently distinct patterns of behaviour. His interest was in the basic problems of evolution and he did a great deal to solve them. He showed how natural selection can determine the fate of a length of DNA, of its bearers, or of a group of relatives.

Many of his ideas were already widely accepted although his work formalised what had sometimes been vague and general speculations. Price also came up with a hint that, under certain conditions, groups whose members are bound together by no more than common purpose (and not necessarily by kinship) may triumph over less cooperative assemblages. A few tricksters might gain a brief advantage in such situations, but an honest society would always prevail at the expense of one in which almost everyone cheats. The actions of the assembly as a whole rather than those of its individual members was what mattered. In such 'group selection' a rising tide floats all boats, however unseaworthy some may be.

For biologists Price's equations deal with material return, measured in terms of numbers of offspring, on an investment of food, courtship or territory. To the intense American they began to have a wider significance. Perhaps, he thought, they could be used to study human society. In that he followed Darwin: 'A tribe including many members who, from possessing in a high degree the spirit of patriotism, fidelity, obedience, courage, and sympathy, were always ready to aid one another, and to sacrifice themselves for the common good, would be victorious over most other tribes; and this would be natural selection.'

Soon Price took that notion further, to suggest that evolutionary rules might be at the foundation of friendship, of family and even of faith. Investment, group membership and deferred reward are, after all, central to all three. He took inspiration not just from *The Origin* but from St Matthew: 'Lay not up for yourselves treasures upon earth, where moth and rust doth corrupt, and where thieves break through and steal: but lay up for yourselves treasures in heaven, where neither moth nor rust doth corrupt, and where thieves do not break through nor steal.' That advice, he began to believe, fitted his own calculations. What mattered was not the trivial prizes available on Earth, but an eternal reward in the afterlife. The Christian Church showed how evolution had made us what we are, for its foundations lay in the Crucifixion, the ultimate act of a divine altruist.

In around 1970, a year or so before I met him, George Price – until then a determined unbeliever – began to persuade himself that the solution for his growing unhappiness was to find salvation in a group of like-minded people. He joined a Christian community and tried to follow its precepts. It was, he decided, prescribed that he must give away his possessions to those less fortunate than himself. They were accepted with alacrity by his new friends among the down-and-outs of Soho Square who, now and again, invaded the Laboratory in the hope of further reward. As his devotion grew and his mind became more disturbed, the middle-aged American began to see divine inspiration in his own career. The probability that his discoveries had emerged by chance was, he calculated, one in ten followed by thirty zeros – a figure so remote that it must be due to more than mere luck. He found messages hidden in the pages of the Testaments and corresponded with creationists even as he delved into the recesses of biological mathematics. As Price put his convictions more and more into practice he gave up his thyroid medicine and ate scarcely at all until, almost out of touch with reality, he came to his tragic end.

At the centre of his models are some very biblical ideas. Ecclesiastes recommends that the faithful should 'Cast thy bread upon the waters: for thou shalt find it after many days.' That statement has messages both literal (invest in stocks against hard times around the corner) and spiritual (subscribe to my faith and cash in with life eternal). To bank managers, bishops and biologists alike, planning for the world to come is essential. For all three, to sin in haste leads to repentance at leisure.

Plenty of creatures plan ahead and do so with, as far as we can tell, no moral judgement at all. The Promised Land flowed with milk and honey, both made in good times as investments for the morrow. Summertime birds, too, lay up stores against the winter. They have prodigious memories. A willow tit stocks seeds in thousands of hideaways and can find them several months later (and it works hard to conceal its caches from pilferers who raid them). Our own ability to do the same keeps tax-accountants in business.

The idea of 'bread upon the waters' is a statement of what biologists call 'reciprocal altruism', a favour done today in the expectation it will be returned in the future. When an ape expends time and energy in scratching another's fur to pick off fleas, it expects to be scratched back with equal or greater enthusiasm. Vampire bats must eat at least once every couple of days or die. When a successful individual gets back to the colony, it vomits up blood to its hungry neighbours who in turn will spew some of the carefully harvested fluid when they themselves have a fruitful night. Many other actions depend on the same logic. Often, males offer food to their females, who provide sex in return. The gift may be expensive for in some insects the female devours him after he has carried out his allotted task but his reward is to pass on his genes.

Price's evolutionary research has been important to students of animal behaviour, but his spiritual ideas, like those of Newton, have had little impact on science. Even so, plenty of scientists and others have tried, with mixed success, to understand how, where, when and why religions have arisen. Nearly all turn, often quite unknowingly, to the ideas of an American mathematician and zealot of forty years ago.

Anything that can be measured is raw material for research,

which means that piety is as open to analysis as is any other endeavour. *Homo sapiens* is, as far as we can tell, the only creature that calls on a transcendental presence to explain the world. Why, when and where did such a personage emerge? The question is not simple. How religious ideas arise and what kinds of societies embrace, or emerge from, belief are related but separate issues, as are the history and structure of each creed and the biological, mental and social conditions that may have produced such groups. Could there be, as George Price thought, a science of religion?

William James certainly thought so. In a 1902 lecture at New College, the divinity school of the University of Edinburgh (an edifice of ineffable grimness later adopted, in a step away from the sublime, as the first home of the Scottish Assembly) he said: 'To the psychologist the religious propensities of man must be at least as interesting as any other of the facts pertaining to his mental constitution.' He was right.

If scientists are to understand something, they must first define it. For chemistry or physics that is straightforward, but for belief the job is less simple. The attempt to do so hints at why believers and non-believers of any faith so often find themselves in conflict. A vast and inconsistent range of convictions is held with equal passion by their adherents; as Herodotus remarked, 'If one were to order all mankind to choose the best set of rules in the world, each group would, after due consideration, choose its own customs.'

Every creed identifies itself as, of its nature, distinct from all others; but in spite of some real contrasts, all the dozen or so major players (together with most of their minor competitors from Anthroposophy to Zoroastrianism) have elements in common. From the bloodthirsty rituals of the Incas to the gentle universe of the Buddha the idea of a force that influences human affairs is universal. The world has gone awry, and the only way to put it right is with a set of transcendent truths. Membership gives privileges denied to outsiders, and adherents are told to defer the pleasures of the flesh for later bliss. Quite often, to join a particular sect involves tests of commitment in terms of expensive or bizarre rituals. To fail the examination means expulsion or worse.

Perhaps the most universal feature of all faiths is a concern with mysteries of their own making. The questions they pose are designed to be insoluble. Their adepts can as a result spend happy hours analysing each other's thoughts with no need for any input from outsiders. Judaism has the secrets of the Ark of the Covenant, Christianity indulges in endless arguments about predestination and free will, Islam is obsessed with the question of the Twelfth Imam who may or may not return, while Buddhists puzzle themselves with the gnomic statements of their founder. A religion without superstitions does not deserve the name. Almost all share the greatest mystery of all: the ultimate cause; the power that lies behind it all, the entity sometimes known as God.

Charles Darwin himself considered that biology had played a part in the emergence of such ideas. In *The Descent of Man* he wrote that: 'As soon as the important faculties of the imagination, wonder, and curiosity, together with some power of reasoning, had become partially developed, man would naturally crave to understand what was passing around him, and would have vaguely speculated on his own existence.' Spirituality is, on that view, a by-product of evolution.

The mechanism behind that involves mutation – the modification of information passed from one generation to the next – together with random changes in populations isolated from each other. Its prime agent is natural selection, inherited differences in reproductive success, which clicks into action as its subjects resist or succumb to disease, to hunger, to sabretoothed tigers or to the disdain of potential bed-mates.

Religions are much the same. Just like species, their success depends on an ability to stay distinct from their rivals. They change with time, in part by chance, but also into better adapted forms which after a time find themselves unable (or unwilling) to exchange information with their neighbours. Differential reproductive success is as central to the evolution of religions as it is to that of life.

Comparative anatomy provides clues about the ancient history of animals and plants, and does the same for every creed. Christianity itself began to evolve into competing sects soon after the Crucifixion. The First Council of Nicaea in the fourth century was called to reconcile incompatible views of the relative position of God and his son. Many Church Councils later, the Great Schism of 1054 split the Eastern from the Western churches on grounds as fundamental as the use of leavened bread in the Eucharist and the nature of the Holy Spirit. Then came the Reformation and its bloody quarrels about salvation by grace, diet, baptism and more. Three hundred years later, the Crimean War was in part sparked off by a row about whether Roman Catholics or the members of the Greek Orthodox Church had the right to the keys of the Church of the Nativity in Bethlehem. Even the placid Church of England is threatened with splits over arguments about female bishops and gay marriage. All these events are recorded in the rituals and beliefs of the competing sects today.

On the larger scale, systems of belief diverge even more. Christians search for a Saviour while Hindus hope for relief from the cycle of life, death and rebirth. Islam depends on obedience to the word of God as dictated to his Prophet. Buddhism sees meditation as the escape from suffering, while Judaism appeals to the rituals of ancient times. The Church of Rome calls on the infallibility of a supreme prelate, but Hinduism has no clear hierarchy of holiness. In Japan, eclecticism rules, for people might go to a Buddhist temple on New Year's Eve, to a Shinto shrine on New Year's Day, and to a Christian church for a wedding.

Diverse as they now are, such systems of thought probably spring, like every plant and animal, from a common source. Many have searched for it, with mixed success and, as is the case in biology, the longer the journey into the past the more ambiguous the results become.

One of the earliest attempts to find the roots of faith was Richard Payne Knight's *A Discourse on the Worship of Priapus* of 1786. Knight was a member of the Society of Dilettanti, a group of aesthetes who travelled much in Italy. There he became interested in the parallels between Roman customs and those of other nations. In his day, the notion of change was beginning to replace that of a universal creator. In the same year as the *Priapus* book, the linguist William Jones set forth a model of language based not on divine intervention at Babel but on the similarities he saw between the extinct Sanskrit written tongue of Northern India and Greek and Latin; 'so strong indeed, that no philologer could examine them all three, without believing them to have sprung from some common source which, perhaps, no longer exists'. Three years later, Erasmus Darwin published his epic poem *The Loves of the Plants* which contained evidence of kinship as manifest in patterns of reproduction. Evolutionary ideas were in the air.

Payne Knight saw that they could be applied to doctrine. He was told of a Sicilian fertility cult in which women who hoped to become pregnant offered wax models of tumescent penises to St Cosimo (some were said to have been given to the British Museum, but they appear to have wilted away). The local priests were happy to profit from the practice although they backed off when the story became public. Knight suggested that all sects, Christianity included, had begun as sexual rites. He compared the Virgin Mary to Diana, the Roman goddess of chastity, and inferred that the faiths of the Old World had, like many of its languages, a tie with India. The Holy Trinity was a descendant of an ancient cult in which the conflict between the active male and the passive female element of a hermaphroditic god created the world and in the end would destroy it. Sexual intercourse was a model of the origin of the universe. He even suggested that the Cross itself was 'the least explicit representation of the male organ of generation'. Outrage at that claim reached such heights that Payne Knight took the unusual step of suppressing his own book.

Provocative as his theories may have been, Richard Payne Knight introduced the idea of a common origin for what had seemed to be distinct sets of certainties. Christianity can chase its own symbols deep into a shared past. The idea of a single god may have emerged when Egypt took up monotheism at the time of Akhenaten's vision of the Sun's disc. A tree, a fruit, a couple and a serpent appear on a Mesopotamian seal and emerge in the Greek Garden of the Hesperides with its golden apples and guardian nymphs. Virgin mothers, too, crop up all over the place. The Sumerian god Tammuz was born of such an individual two thousand years before Christ, and the Church of the Nativity in Bethlehem is said to be on the site of his shrine. Other pregnant virgins include Isis, mother of Horus, and Juno, mother of Mars.

For gods, death and resurrection is an occupational hazard. Tammuz suffered the experience as did his Greek equivalent Adonis while Osiris, husband of Isis, was allowed to return from beyond the grave to fertilise his wife. The Greeks had the resurrected Dionysus, god of wine, while the sun god Ra, Baal, Dionysus, Jesus and many more had the same experience, as proof of their divine power. Many of the most potent images of Christianity have ancient roots and many have tried to pursue them through archaeology, myth, psychoanalysis and more.

James Frazer in his 1890 work *The Golden Bough* followed Payne Knight in the claim that the Mediterranean creeds began as fertility rites, but for him they involved crops rather than people. Ceremonies marked the death of plants and animals in the dry season while a few months later their return was celebrated as the rain consecrated its marriage with the Earth. In ancient times, a god-king was sacrificed each year and was resurrected in a new guise. A farewell supper for the victim was held by his devotees and after the ceremonial execution of their leader the members of his sect saw themselves – unlike their unfortunate ruler – as 'born again'.

Frazer may have been right about recent history but the true sources of Christianity and its close relatives must be much older. Recent discoveries hint that an ancient near-Eastern cult played a part in the origin of farming, rather than modern denominations springing from the soil of the first fields.

In the 1990s a Kurdish shepherd noticed some square blocks on the summit of a mountain called Göbekli Tepe – Potbelly Hill – in southern Turkey. Archaeologists uncovered a remarkable sight: the oldest buildings ever found, constructed eleven thousand years ago, at (or perhaps even before) the very beginning of agriculture. The ruins date from eight and a half thousand years before the Giza Pyramid and eight thousand years before Stonehenge. Potbelly Hill is the most important such site in the world. Just four of its twenty circular structures, each up to thirty metres across, have so far been excavated. Each is held up by limestone pillars two and a half metres high. They are decorated with reliefs of lions, boars, snakes, vultures, crayfish and other creatures, together with a variety of abstract designs. A figure of a headless corpse picked at by vultures hints at the existence of a culture of 'sky-burial' like that of today's Zoroastrians, in which the birds take the deceased to an afterlife high above.

The place was built by tribes who hunted game and gathered fruits and seeds, although some may just have begun to cultivate semi-wild grasses and to settle into small villages. The countryside, now a parched and stony wasteland, was then lush and green. It teemed with wildlife and hundreds of thousands of bones of gazelles and wild cattle have been found nearby. The most extraordinary aspect of the complex is that quite soon after it was built it was buried, to add a further cost in labour to an already huge investment.

Why did these people – seen by Frazer and others as possessed of little spiritual insight – build such a place? Before the discovery of Göbekli Tepe the standard story was that life, and the imagination, could not be liberated until after the move to the land, when cultivation brought abundance and led to introspection, to intellect and to advanced systems of devotion. As production and populations grew, men and women had to find rules that allowed them to cope with life in larger groups. For that, the notion of a shared deity was a great help.

The Turkish ruins tell a more ambiguous tale for they hint that such a doctrine began before farming and its population explosion took much of a hold. No remains of houses have been found close by so that the monument may have been a place of pilgrimage; a prehistoric Chartres Cathedral that attracted visitors from miles around. So many arrived that in time wild game and wild plants ran out. Its guardians were forced to cultivate more and different plants for food – and the ancestors of modern wheat, rye and oats were indeed domesticated nearby. God enjoined that Adam, the first farmer, would have to work hard to feed himself: 'In the sweat of thy face shalt thou eat bread.' A little later, when his son Cain was exiled for having killed his brother Abel, he 'went out from the presence of the Lord, and dwelt in the land of Nod, on the east of Eden', which may be near Göbekli Tepe. A statistical study of language suggests the Indo-European tongues identified by William Jones emerged in just that region at about the same time, as further evidence of its cultural importance. Perhaps the huge increase in numbers that took place, and the decrease in health and social stability that came with it, led the temple's disillusioned devotees to bury it as a place of evil, best forgotten. It was lost for eight thousand years.

Ideas of supernatural life, death and rebirth must be older even than that great ruin, for they are implanted across the globe in places that did not come into contact until recent times. Some may come from the natural world, from the phases of the Moon or from the tilt in our planet that causes shifts in the lengths of each day. At the winter solstice the Sun reverses its daily movements in the point where it sets and begins to climb in the sky, no doubt because its deity has died and been reborn. From Stonehenge (which is aligned with that point on the shortest day), to Amaterasu's escape from her midwinter cave in Japan, to the drunken Roman Brumalia on December 24th, the Dongzhi celebration of China on the 21st as yang (light) begins to triumph over yin (darkness) and to the Maori ritual of Maruaroa o Takurua held on the southern winter solstice of June 21st, the idea of rebirth is set into the mind by the facts of physics.

Quite how much further into the past such notions extend

is hard to tell. Some see hints of mysticism in the worked arte-facts of a million years ago, a few of which are said, with a certain optimism, to be fashioned to look like heads. The first graves, and the first hints of another world, come from ninety thousand years ago in Israel. Seventy-five thousand years before the present the inhabitants of a South African cave cut geometric patterns on pieces of red ochre, perhaps with a spiritual motive.

Forty thousand years ago came some real steps towards spiritual practice. Musical instruments made of vulture wing-bones appear in caves in Germany, as does a statuette of a creature half man, half lion and the first of many Venus fig-urines, whose exaggerated bodies suggest a mythic role. Soon afterwards rock art emerged in Europe, in Namibia and in Australia. Such changes might reflect a social advance, perhaps the origin of true language. An ability to preach would be a great help to those who hoped to share an explanation of what they see around them.

The study of the beginnings of religion is rather like that of the origin of life, for the subject is filled with guesswork. Should it be classed as biology or social science? Did evolution form our beliefs, or did the idea of the supernatural help make us human? It may not be easy to tell.

Mountains of words have been written on the topic, to no clear end. David Hume's 1757 book *The Natural History of Religion* speaks of the 'tendency among mankind to conceive all beings like themselves, and to transfer to every object, those qualities with which they are familiarly acquainted, and of which they are intimately conscious'. God is no more than an exaggerated father figure, a ghostly, morally perfect and omnipresent version of the powerful parent who raised us. He can explain what is around us and tell us what is to come. Our curiosity about the world generates the sense that every action must have a cause (Hume: 'A purpose, an intention, a design strikes everywhere even the careless, the most stupid thinker') and who else could be responsible but a deity? Our pre-human ancestors who turned stones into tools must have noticed that while they made axes, axes did not make them. One was made and the other was the maker – but who formed the creature that made the tools?

Man's fascination with cause led, in the end, to science. When that is denied, or fails to provide an explanation, or when – as is always the case – its account is incomplete, a penchant for the supernatural is almost built in. Doubt rather than certainty is what demands an effort of will.

The journey between the two can be seen in every child. The young often have imaginary friends or ideas about death-defying heroes. They tend to accept what they are told, however unlikely. The passage from infant credulity to cynical adulthood suggests that belief may be as implanted in our frames as are the many other attributes that shift at adolescence. If so, the foundations of faith must be laid down long before the intellect develops. Not for nothing do the Jesuits say, 'Give us a child until he is seven and I will give you the man.'

Children two years beyond that crucial birthday were set to play a staged game. They had to choose which one of two boxes held a reward, to point at it, and then to open it. They were told that they were in the presence of an invisible authority (called, for some reason, 'Princess Alice'), who would make a sign if they identified the wrong one. When they made an incorrect choice the experimenter furtively made a light flicker. To an infant, the subjects shifted their bet; they accepted, without question, the presence of a higher power.

In another unkind experiment, boys and girls were persuaded that a scientist had invented an apparatus that made a duplicate of whatever was placed inside it. Each child's favourite stuffed animal was put within, the lights flashed and the machinery hummed, and with some sleight of hand the child was given the toy back again and told it was a replica. Almost without exception, he or she rejected it. Somehow, the supposed copy had lost a mysterious quality present within the original. For the tearful subjects, reality involved more than the real.

Adults can be just as gullible. A Boston Red Sox supporter once buried a team jersey in the foundations of Yankee Stadium in an attempt to damage their prospects. The Yankees paid fifty thousand dollars to have it dug out. Even on this more sceptical side of the Atlantic there was an uproar when Redditch Borough Council approved an ecologically, if not theologically, sound plan to warm up the municipal baths with waste heat from a crematorium (they went ahead anyway). In a more serious illustration of the seductive power of essence many people are dismayed by the thought that their organs might be used after their death in transplants or in medical research.

In a further test of the juvenile psyche, psychologists played a crafty trick on a group of four-year-olds. Each child was given a marshmallow and told that if they waited for twenty minutes before they ate the treat, they would be given another. Most cracked within seconds and gobbled the titbit, but about a third held out and – with smug expressions, no doubt – got a second confection when the experimenter returned. Many more were able to resist the temptation when he suggested diversions such as singing a song, reciting a nursery rhyme, covering their eyes, hiding under the table, or rocking back and forth. Ritualised behaviour fortified their resolve.

The elements of devotion – a mysterious power, a hidden essence, and a fondness for ritual – are all present in the unformed mind. From there to Jerusalem's Temple, the Vatican, or the Dome of the Rock the rest may be detail.

Another approach to the biology of faith is to ask questions like those used to investigate height, weight, sport, intelligence, or crime: does the attribute run in families and how much of that is due to shared culture and how much, if any, to the double helix? If genes are involved can we find them? And what of the structure that generates the sense of devotion; where within the grey matter does it lie? Are there individual differences in brain activity that make the supernatural more attractive to some and, if so, why?

Responses to statements such as 'I am certain God exists', 'I pray every day', 'My holy book is true' or 'Mine is the one true religion' can sort out the sanctified sheep from the sceptical goats. Identical twins are more similar to each other in such measures than are those who share no more than half their DNA, and – as is the case for IQ – that similarity increases when they leave home. In athletics and horse-racing, the importance of genes varies with circumstances and the same is true of belief. In the general population, about a fifth of the total variation in the readiness to accept a higher power reflects genetic diversity, but for fundamentalist families, the Olympic champions of the religiosity stakes, the figure is twice as high.

Other insights come, again as in sport, from the machinery of the body. Neuroscientists (themselves sometimes less sceptical than they might be) have scanned the brains of devout subjects as they think elevated thoughts. It is hard to define just what a mystical insight might be, let alone to summon one up on demand, and the results are equivocal. Some say that the spirit spot is on the right (and emotional) side of the skull, others that God is on the rational and analytical left. Enthusiasts are certain that they have found separate areas that light up when a subject experiences an intimate relationship with the deity (right), fear of his power (left), and doubts about his existence (right again). A tiny section called the anterior insular cortex lights up when mothers look at their children or when people see a smile. This magic spot also responds, some say, when its owner is in 'a state of union with the divine'. Perhaps that is the home of the other world; although a more reasonable interpretation is that such a subtle attribute involves much of the conscious brain and that today's technology is not good enough to find it.

Medicine has often used the abnormal as the key to the normal, for symptoms of disease may give a valuable insight into the machinery of the healthy body. Perhaps the same is true of the mind.

All creeds, of their nature, involve social activities. They depend on the ability to read the feelings of other believers, whose shared emotions help form the experience, and, most of all, to know the mind of God. The talent starts early for if a baby points at an object, it must realise that its mother is paying attention. It makes leaps and bounds with every year.

Adults, too, can be tested for empathy, for how well they appreciate the moods of others. They are shown pictures of faces manipulated to produce slight smiles or frowns and told to identify whether the person is terrified, amused, regretful or flirtatious. They must also respond to statements such as 'I find it hard to sustain a conversation' or 'I prefer to go to a museum than a theatre' to test their preference for solitary or for communal pastimes. The ability to spot prime numbers, or word patterns, is another pointer to their interest in the inanimate as opposed to the thoughts of their fellows.

Men, on average, are worse at identifying moods but better at prime numbers (an ability that demands no insight into anyone's emotional state) than women, while university professors as a whole are more detached than average from those around them, with scientists the most self-centred of all. When people are asked about their sense of devotion, women are at the top, and scientists at the bottom, of that list too.

One condition leads to a life almost cut off from society. People with autism lack empathy, concentrate on themselves alone and may be obsessed with a single activity. Those with the milder version of the condition known as Asperger's syndrome are clumsy, shy and tongue-tied but most of the time are able to cope. Others on the spectrum do much better, for they have 'high-functioning autism'. Such people are successful and effective, but have little sense of anyone's emotional state and often show a deep interest in matters mechanical and numerical. A civilisation run by autists (and perhaps even by scientists) would collapse at once.

Autism and its relatives are much more common among men than women and, at least in its severe forms, the condition is highly heritable. On the emotion tests, those with autism proper do worst, then Asperger's people, followed by the high-functioning group, and then in order by mathematicians, scientists, university professors and male members of the general population. Women do best of all.

Those with autism or Asperger's dislike notions that are not supported by tangible evidence. They are just half as likely to express trust in God as average. Members of the high-functioning group are also far more willing to class themselves as atheists or agnostics than are their fellows. In decreasing order of scepticism come people with autism, Asperger's patients, scientists, university professors, men as a whole and women in general. A scale of the ability to read another's mind is hence a mirror image of the tendency to believe.

Perhaps an inborn logical, systematic and self-centred personality (that of autists, scientists, and males not excluded) disposes to doubt, while those blessed with a more responsive frame of mind find it easier to summon up the divine. The more sympathetic members of society feel in emotional contact with a deity while those trapped in their own mental universe are happy with their own thoughts.

Given such inbuilt variation in the willingness to accept mystical ideas, how did such notions become so embedded into society and why have they almost disappeared from some places even as they flourish in others?

The rise and fall of any system of belief is, like that of all other institutions, a messy business. It has, no doubt, a multitude of causes and is driven by forces that change over the years and from place to place. George Price was inspired by the Crucifixion as sacrifice, and by the notion that the promise of eternal life required believers to show altruism in the Earthly sphere. He could, and did, point at the fact that many of today's great hospitals, schools and charities were ecclesiastical foundations and that many church-goers are decent, honest and generous people. For him, and for many believers, piety is the glue that holds society together.

Others are less magnanimous. They note that the equations for altruism can be used, with minor modification, to understand selfishness, retaliation and revenge. To their minds, religious doctrine is a conspiracy of the powerful against the weak, concealed under a thin cloak of generosity. Its practices are no more than a cynical attempt to disguise conflict and greed as cooperation and kindness.

Proponents of either view can each appeal to the Bible for support, as long as they choose the right verses. The heroes of the Pentateuch – of Genesis, Exodus, Leviticus, Numbers and Deuteronomy – were the people of an unforgiving god, aided by a powerful priesthood, who kept them on the straight and narrow by punishing anyone who fell away from his precepts. They had a great interest in the settlement of scores: 'Eye for eye, tooth for tooth, hand for hand, foot for foot, burning for burning, wound for wound, stripe for stripe' and even their deity liked to get his own back: 'I will make mine arrows drunk with blood, and my sword shall devour flesh ... Vengeance is mine; I will repay ...' Huge differences in wealth are a cause for celebration: 'Now the weight of gold that came to Solomon in one year was six hundred threescore and six talents of gold.' His wealth was used in ostentatious celebration of the power of the Lord. Much of the Old Testament tale unfolds against a history of mass murder, of sexual violence and of slavery.

It reflects the society of the time. Excavations at Ur from around 2500 BC – a few centuries before Abraham himself was born there – show that human sacrifice was common: near a royal corpse are seventy skeletons of soldiers and courtiers, clubbed to death and embalmed in salts of mercury to accompany the ruler to the next world. Most of the faiths of those years were violent, greedy and exclusive.

Later sections of the Old Testament begin to hint at a less flinty universe, as when Elijah rewards a widow who offers him her last few grains of barley and drops of oil by providing her with a long-lasting supply of both and cures her dying son. Even so, the New Testament is far more concerned with clemency and fairness (as in the words of Jesus to a crowd about to stone a woman to death: 'He that is without sin among you, let him first cast a stone at her'). It is, as a result, much appealed to by those with a positive attitude to religion. Mutual aid is at its heart; as Luke puts it: 'And as ye would that men should do to you, do ye also to them likewise.' It speaks of a wish to welcome everyone into a universal embrace and of an era of peace on Earth. The affluent are warned that wealth is a burden rather than the key to paradise ('Verily I say unto you, that a rich man shall hardly enter into the kingdom of heaven'). Punishment comes in now and again (and in its last chapter all hell breaks loose) but self-sacrifice and cooperation are at the centre of its message. The New Testament is set firmly in the modern world, with unselfish love for one's fellows at its centre.

The Good Book's many verses hence provide a defence for quite contradictory models of the evolution of faith. The positive image of the pious and the darker notions of those who doubt can be tested against the facts of history. Each side can call on evidence that supports their own view.

To Christians, the succinct verse of Ecclesiastes that 'if two lie together, then they have heat: but how can one be warm alone?' tells of the power of piety. Shared devotion is essential to society and only with its emergence could a harmonious structure evolve. Clergymen are the equivalents of social workers, while godliness provides a moral framework that allows culture to flourish. Perhaps, in ancient times, devout groups were more stable than gangs of quarrelsome doubters; a suggestion close to George Price's claim that cooperative groups of animals may prevail over others.

The role of such 'group selection' in the natural world is far from clear, but the patterns of growth of creeds and

churches fit the notion rather well, perhaps because ideas per-
colate far faster than genes. Faith flourishes through the
survival of the nicest. The generous prosper, reciprocity
spreads and in time a vague and benignant piety takes hold.
Cultures in which each member wants only to get the best for
himself do not survive for long.

The cohesive power of shared belief can be seen in the fate
of the hundreds of Utopian communities that appeared in the
United States in the nineteenth century. Some were Christian
in character, while others had no more than a political or economic agenda. The Oneida Community began in 1848 in
upstate New York. It was based on the premise that Jesus had
returned long before and that it was time to build a heaven
upon Earth. Its adherents shared their possessions and listened
to lengthy sermons every day. Each member was subject to
constant criticism, was obliged to live an austere life, and was
discouraged from contact with the outside world. Every man
was, in principle, married to every woman ('male continence',
sex without ejaculation, prevented too much illegitimacy).
Births were arranged on eugenic grounds and many of the
children were fathered by God's representative, the colony's
founder. Babies were raised communally, with few parental
visits. The sect lasted for forty years, and its commercial branch
(which makes cutlery) still exists.

The town of Harmony in Indiana, now a tourist trap, is a
microcosm of how religious certainties help to keep communities in being. The Harmony Society of Indiana was
founded in 1814 by German Lutherans. Their rigid tenets led
to real progress: 'The settlement made more rapid advances in

wealth and prosperity, than any equal body of men in the world at any period of time, more, in one year, than other parts of the United States ... have done in ten.' A decade later the Lutherans sold up and moved to a better site. The town was bought by the Welsh socialist Robert Owen who aimed to set up a secular commune that had political goals, but no spiritual element. New Harmony, as he named it, was based on the hope that joint labour and freedom of speech would ensure a terrestrial paradise, with no need for a saviour. Quite soon, the community was plagued by people who joined but did not join in. It crashed within two years and the place reinvented itself as a normal township with somewhat left-wing views.

The fate of Harmony, Old and New, was reflected across the nation. Almost without exception America's Christian communes lasted several times longer than did their humanist alternatives, and the more assertive the celebration of their creed the better they survived. A successful community needs to show the world its power, and to make it clear to possible recruits that they can gain from being part of it, which does a lot to explain Chartres Cathedral and its equivalents. Without shared belief that edifice – and society itself – could never have been built.

All political parties and all nations, like all religions, know that shared sentiments are a force for unity – but how can they be identified and, with luck, fostered? Cooperation, spirituality and even love could be added to Darwin's 'patriotism, fidelity, obedience, courage and sympathy' as elements of the social glue, but such things are hard to pin down. Another emotion is more palpable and is at the centre of many faiths. Grief is immediate, devastating and easy to recognise. As Dr Johnson put it: '... for sorrow there is no remedy provided by nature. It requires what it cannot hope ... that the dead should return or the past should be recalled.' Most people do recover in time, but some face prolonged desolation. They pay a price in both mind and body. Many children whose fathers were killed in the Kosovo War still suffered severe depression a decade later. An increase in the stress hormone cortisol suppresses the immune system and leads to an increased chance of illness. People who lose a partner double their own chances of mortality – and for young men that figure doubles again.

Not long ago I myself saw the power of sorrow as a social adhesive. I lived at the time next door to the celebrated singer Amy Winehouse, who died in 2011 at the age of twenty-seven from a huge overdose of alcohol. I came home to find the street blocked off with police tape, and a group of weeping friends and relatives behind it.

Next day fifty people came to pay their respects, then a hundred, then more – a cross-section of London in all its diversity, together with many from Europe and even some who had come from South America. Often, those around the impromptu shrine exchanged sympathies and even tears with strangers. More than a year later, mourners still appeared every day and there have been thousands since the event itself.

They left flowers, messages and bottles of vodka as tributes, matched by votive candles and crosses (Ms Winehouse was Jewish). Their epistles were poignant and asked that Amy rest in peace or be happy in heaven. But why should a great mass of people, united by no more than esteem for a particular singer, invest so much energy in somebody they had never met? The power of the group is at work. Sorrow binds people together and assures the bereaved that society shares and dilutes their pain. Even in a secular nation such sentiments are still a force for unity.

The Church has not been slow to notice. Grief and the recovery therefrom are at the heart of the Christian message. Easter – the experience of death and rebirth – is a high point of its calendar. Many of its rituals are related to mourning, which is among the few contacts that it still has with the British people as a whole. Unbeliever as I am, I found the humanist services that accompanied the funerals of each of my parents oddly unsatisfying, for they lacked the familiar and eloquent rites that bind those who remain with the departed and with each other. Even after death, belief is a force for unity.

Sceptics have a rather different view of the origin of religion. Its rituals may appear to bring people together, but whatever their surface attractions, in its fundamentals religion emerges not from the benevolent rules of the New Testament, but from the violent and turbulent world of the Old. Karl Marx's comment on dogma as a drug is familiar, but the full quotation reveals a deeper insight into its roots: 'Religious suffering is, at one and the same time, the expression of real suffering and a protest against real suffering. Religion is the sigh of the oppressed creature, the heart of a heartless world, and the soul of soulless conditions. It is the opium of the people.'

That narcotic, he thought, had nothing to do with self-sacrifice and mutual aid. Instead, the priesthood was a tool of repression used by rulers to preserve their station. Clerics were armed with untruths (although swords were available when called for). The weak are kept in place with the promise of happiness after death, miserable as their Earthly lives might be. The imagined deity is no more than a secret, albeit non-existent, policeman.

The idea of belief as an agent of social control goes back to long before *Das Kapital*. Glaucon, in his discussions with Socrates in Plato's *Republic*, insisted that people would behave well only if they thought they were being watched. Even the most honest person given a ring of invisibility would act like a knave. A hidden eye is a necessary fiction, a falsehood that keeps the peasants in their place; as Gibbon put it in *The History of the Decline and Fall of the Roman Empire*: 'The various forms of worship, which prevailed in the Roman World, were all considered by the people to be equally true, by the philosopher as equally false, and by the magistrate as equally useful.' What spiritual ideas took hold was unimportant, as all that mattered was that the masses subscribed to one of them.

William James felt much the same. Flawed as any dogma might be, it persuades its followers to 'act *as if* there were a God; feel *as if* we were free; consider Nature *as if* she were full of special designs; lay plans *as if* we were to be immortal; . . . we find then that these words do make a genuine difference in our moral life.' If it succeeded in that, its doctrines must in some sense be true and even if the law-makers knew that they were not, those who accepted them would improve their behaviour to match. To belong to a church was to live a lie, but as long as those in charge kept that to themselves, they – and the people they ruled over – would benefit.

Sects tend to appear at times of uncertainty; as David Hume put it: 'All human life, especially before the institution of order and good government, being subject to fortuitous accidents, it is natural that superstition should prevail everywhere in barbarous ages, and put men on the most earnest enquiry concerning those invisible powers who dispose of their happiness or misery.' He was right. In North America, the Ghost Dance movement reached its peak in the 1880s. In despair at the loss of their land Native American seers preached that a ritual dance, which lasted for days and led its participants into frenzy, would cause a new skin to spread over the Earth. It would bury the colonisers and restore the happier times of old. The idea spread across the Great Plains and was ended by the massacre of the Sioux at Wounded Knee.

Many other revivals have taken place in periods of upheaval. In the United States at the time of the Great Depression, the Southern Baptists and the Seventh-Day Adventists had a surge in membership at the expense of more liberal denominations. Within that nation, patterns of inequality and of adherence to Christianity from the 1950s to the present show, again and again, that the second follows the first; that a period of injustice and social pressure leads to a rise in the numbers of church-goers. As the proletariat's position declines, the attractions of the pews increase – which may be why in these straitened times, from the Bible Belt to the Arab

World and from Africa to Indonesia, fundamentalism is once more on the rise.

As Marx might also have predicted, across the world there is a precise fit between social unfairness and the power of the priesthood. In countries whose governments are fair and effective, the influence of the clergy fades. The most devout nations have more crime, more infant deaths, more mental illness and less social mobility than do those in which dogma plays a lesser part. Their citizens trust their God more and each other less and the Churches gain as a result. Chaos and credulity go together.

Marx would also be gratified to learn that within almost all states, the rich are more devout than are the poor. In some, the pious became wealthy: Gibbon notes a Benedictine abbot's claim that 'My vow of poverty has given me a hundred thousand crowns a year; my vow of obedience has raised me to the rank of a sovereign prince' (Gibbon claims to have forgotten the consequence of the abbot's vow of chastity). It seems that, rather than the weak turning to the church for comfort, the powerful, consciously or otherwise, foment its efforts to divert the lower classes and to keep them in their place. Armed with such observations, the Soho philosopher can rest easy in his Highgate grave.

As the ruins at Göbekli Tepe hint, it all began with capital. Napoleon's succinct view was that 'Religion is what keeps the poor from murdering the rich.' Karl Marx took that radical idea further. He talked of the 'primitive communism' of hunter-gatherer societies, in which there was no private property, few political entities and little need for social control.

With farms and their surplus value came wealth. The invention of the invisible chief constable ensured that it could be gathered into a few hands, and that it stayed there. Tablets dating from around 3000 BC in Iran suggest that at the beginning of Old Testament times workers were already seen as no more valuable than cattle, while their rulers had names such as 'Mr One Hundred', which reflected the numbers of people below them. The plebs ate porridge while their rulers gorged on meat.

Modern hunter-gatherer societies are still rather equitable, with a distribution of assets not much different from that in social-democratic states such as Sweden. Their members do not allow anyone to have too much power. Their devotional practices, such as they are, turn on individual belief and not on hierarchy and show. In stark contrast, today's peasant societies have differences in relative affluence as great as in the United States. As the global economy shifted towards farming, and as inequality grew, statements of influence and piety, from Göbekli Tepe to Seville Cathedral, were built to remind the populace of the great detective in the sky.

The imagined deity and his police force have always been swift to act against those who break the rulers' rules. The Book of Leviticus, with its brutal penalties for those who do not conform to arbitrary laws on food, sex and ritual, shows the process at work, while the Inquisition brought the practice closer to the modern world. Another Marx, Groucho, pointed out the difficulties faced by any faith when he said that 'The secret of life is honesty and fair dealing: if you can fake that you've got it made.' If a badge of devotion is expensive, those who are not willing to invest in it can be identified and punished or expelled. None of this fits with the notion of shared conviction as the main ingredient of a harmonious and equal society.

Which of those world-views best explains the emergence of Christianity, or of any other faith? Their appeal – altruism and rectitude as against greed and trickery – depends on the observer's point of view. For enthusiasts for any dogma, its certainties, its members and its priesthood seem noble, while to outsiders they can appear absurd, vicious or self-defeating. No doubt the acolytes of the Inca priesthood who fattened up children before pulling the living hearts from their chests, and the parents who welcomed their screaming infants' tears as signs that the rains would soon come saw their spiritual leaders as upright servants of a higher power, just as do those who pack the pews of fundamentalist churches in the American South even as their pastors embezzle millions. There are on the other hand undoubted heroes in Christianity and other creeds, with noble figures such as Father Damien de Veuster, the Belgian priest who assisted the lepers abandoned on the Hawaiian island of Molokai before himself dying of the disease (he is now Patron Saint of the Diocese of Honolulu and Hawaii).

Arguments about the merits or otherwise of belief (let alone of particular faiths) tend to degenerate into slanging matches, with each side picking facts to fit its case, although it must be said that at least in the modern world the facts seem increasingly to support the sceptics. Rather than speculating about the intellectual or social forces behind the success or failure of any religion, it may be easier to disentangle them in Darwinian terms.

Throughout history the fate of each canon has depended on the force that drives evolution; on the ability of one lineage to copy itself faster than another. Natural selection on belief is still hard at work, as evidence that faith is, in many ways, an evolved part of the human condition.

The Bible has a deep interest in its subjects' reproductive lives. The first of six hundred direct instructions from God to his people in the Old Testament is 'Be fruitful and multiply.' Many of the adherents of particular creeds obeyed that instruction with enthusiasm and enjoyed spectacular success as a result.

The Amish broke off from the Anabaptist movement and migrated to that New Jerusalem, the United States of America. There they split again, into a strict 'Old Order', and a less rigid group. The former group lived – and live – a rural life based more on the eighteenth century than the twenty-first. Most of its members have no interest in education and almost all marry within the community.

In its early days large numbers of children were born, but almost as many died, so that in 1900 the group had no more than about five thousand members. Quietly, they kept their high fecundity even as infant survival improved and the numbers born to their fellow-citizens decreased. Today, each mother has, on average, half a dozen children and the community is growing at almost 10 per cent a year, which means that numbers double every fifteen years or so. At that rate the Amish could, in principle, by the middle of the next century, have a population equivalent to that of today's United States. Those who leave to live secular lives have fertility not much higher than the American average (and their brethren in Germany have faded away since they turned, like their fellow-citizens, to the pill).

Few can match such sexual zeal, but across the entire globe the devout, of whatever denomination, are almost always more fertile than their secular neighbours. Taking information from every major creed, the most pious average two and a half children, those who go to a religious ceremony no more than once a month exactly two, while those who pay no attention to their national orthodoxy end up with just one and two-thirds offspring. Switzerland is a microcosm of the contraceptive effects of atheism. There, Hindus manage 2.8 births per woman, Muslims 2.4, and Jews 2.1. Christians themselves produce no more than 1.5 but non-believing Swiss women do even worse, for on average they bear just 1.1 children. Such patterns mean that almost no primarily secular nation – and that includes most of Europe and almost all Britain's former Dominions – is now able to maintain its population through the reproductive efforts of its own citizens, although some compensate through the higher fecundity of (often faith-based) immigrant groups.

Priests have long understood the power of numbers. Exodus describes the Egyptians' robust approach to the growth of the numbers of outsiders: 'Come on, let us deal wisely with them; lest they multiply, and it come to pass, that, when there falleth out any war, they join also unto our enemies, and fight against us, and so get them up out of the land.' Judaic texts are filled with exhortations to be fruitful. Rachel, wife of Jacob, in despair at her sterile state, says, 'Give me children, or else I die' while Deuteronomy glories in the fact that, for the Jews, 'The Lord your God hath multiplied you, and, behold, ye are this day as the stars of heaven for multitude.'

Israel itself now faces the ancient Egyptians' problem. It has almost eight million citizens. In the 1990s, the numbers of Jews were boosted by the arrival from the former Soviet Union of around a million who claimed that identity. Now the balance has shifted. In the 1950s, nine-tenths of its inhabitants were Jewish, but because of the increase in the Arab population that proportion has dropped to just three in four. The number of children per woman is almost four for Israeli Arabs and fewer than three for Jews over all. If present patterns continue Israel will soon have to choose whether to remain a Jewish state or face the uncomfortable fact that in their democracy non-Jews will make up the largest bloc of voters.

That dilemma has a message for every religion; that when it comes to evolution, be it of flesh or of faith, demography is all.

Christianity owed its early success to the processes outlined by the Reverend Thomas Malthus, the proponent of the importance of population growth when faced with limited resources, whose book *An Essay on the Principle of Population* gave Charles Darwin the idea of evolution by natural selection. Its future may be driven by the same rules.

The Acts of the Apostles speak of mass conversions of several thousand souls at once, but such unlikely events are not needed to explain how the sect grew so large within no more than fifty years of the Crucifixion. What was at first a tiny group of disciples expanded at around 3 per cent a year. That is about the rate at which the Mormon Church has grown since its foundation in 1830. The sexual appetites of the followers of Joseph Smith were crucial. The Utah settlers averaged almost nine children each and some mothers had twenty. Their numbers exploded.

Early Christians were the same. Pagan Rome, militarised as it was, had become a masculinised society. Women were much less important than men: they could not vote or hold office and were, in effect, the property of their fathers. Many girl babies were killed. In a letter from around the time of Christ a Roman citizen wrote to his wife: 'I am still in Alexandria . . . I beg and plead with you to take care of our little child, and as soon as we receive wages, I will send them to you. In the meantime, if (good fortune to you!) you give birth, if it is a boy, let it live; if it is a girl, expose it.' Inscriptions at Delphi show that of six hundred families, some with many children, just six raised more than a single daughter. The result was inevitable. A contemporary historian blamed the decay of the Empire on a shortage of females, with in Italy as many as a hundred and forty men for every hundred women.

Christianity, in contrast, forbade abortion and infanticide and was strict in its injunctions against divorce, polygamy and incest, all of which flourished in the pagan world. It even collected up the baby girls left to die of exposure, and gave women high status, both in civil life and within the primitive church. This attracted members and increased the numbers of infants born ready to be received. Many converts were well off and as early as AD 57 Tacitus writes that members of the senatorial class were at risk of 'foreign superstitions'. Such people had more surviving children than others.

Inevitably, some of the lonely men who roamed the streets of Rome married Christianised women, which added to the faith's demographic power. As St Paul put it in his letter to the Corinthians: 'The unbelieving husband is sanctified by the wife' and full conversion was the obvious next step. Later emperors became alarmed at the throng and forbade missionaries to call upon women, but many disobeyed. Female desire to accept spiritual comfort is still powerful, for women were the backbone of charismatic movements such as the Shakers while most of the converts to Protestant fundamentalism in today's South America are of that gender.

The triumph of Christianity, now the world's largest faith, was from its earliest days driven by biology – the survival of the sexiest – as much as by belief. The process has been renewed. Malthus argued that 'moral restraint' rather than 'the vices of mankind' such as hunger, disease and war should be used to limit numbers. Although they use birth- rather than self-control, plenty of places have taken his advice, with small families, started late in life. Portugal, Italy, Latvia, Japan, Korea and many other nations are now well below replacement level.

Elsewhere, 'moral restraint' has had less impact. That demographic imbalance will, as it did in Christianity's first days, have dramatic effects on the future of religion, in the near term at least.

It was once assumed that black Africans and Europeans sprang from separate roots. Absurd as that notion might be, it is based on the undoubted fact that they do look distinct.

Shifts in the incidence of the genes responsible for skin pigmentation are a classic example of Darwinian natural selection. They also show how group differences in reproductive success affect both body and mind.

The evolutionary part of the story has to do with the demand for vitamin D (a shortage of which causes rickets, multiple sclerosis, and more). It can be made through the action of ultraviolet light on certain body constituents. Our distant ancestors were black to protect themselves from the effects of skin cancer and of sunburn, but their native continent had plenty of sunlight, whose rays could penetrate far enough into the skin to do their health-giving job. On the move into the cloudy climates of the north, the balance shifted and vitamin D ran short. New mutations appeared that reduced the amount of melanin. They were favoured because they enabled people to make the crucial chemical, which was (and still is) in short supply in such places.

In more recent times patterns of pigmentation have changed fast for reasons that have nothing to do with vitamins. For several centuries after 1492, the global abundance of the genes for whiteness shot up as Europeans set out in their millions to colonise the world. In the Americas in particular pale skin became more and more common as settlers streamed across the Atlantic. Soon, the balance shifted again as vast numbers of African slaves were imported. In time, all parties exchanged DNA, which means that the evolution of the continent's modern inhabitants has been driven much more by differences in the demography of its founding groups rather than by natural selection.

Columbus' whitening world is now in full retreat. Genes for melanin pigment are booming as never before. Niger has a birth rate five times higher than that of most of Europe. In 1950 there were about twice as many whites as blacks in the world but nowadays the figures are about equal. If present trends continue, in 2050 explosive growth in Africa coupled with stability or decline elsewhere will result in twice as many blacks as whites.

Africa's expansion comes from its inhabitants' reluctance to limit their families (often on the advice of their pastors). The shift in the world's faces hence has nothing to do with the biology of melanin; the shared behaviour of those who are blessed with lots of it, everything.

The triumph of Africa will alter not just the world's genes, but its minds. In almost all developed nations Christianity has lost the demographic war. Even in the United States the proportion of adherents is in decline, for the number who state their affiliation to religion as 'none' has doubled in the past two decades to almost one in five. Congregations in the once devout lands of Austria, Ireland and Switzerland have collapsed. More than half of all Czechs consider themselves to be atheists or agnostics while in Britain the Welsh take the pagan crown. The 2011 census records a fall of more than 10 per cent in the proportion of Christians in just a decade.

St Paul's Cathedral and its fellows, once their local spiritual and economic centres, have become tourist traps, admired for their beauty rather than their meaning. Much the same has happened to the words preached from their pulpits. In Europe at least, religion is in its twilight years. Quite soon, many churches will find themselves trapped in the predicament of minority languages; that as the numbers of speakers go down, their culture becomes less and less attractive. In several countries Christianity may in effect disappear, perhaps to be replaced by something else.

Many are concerned that in Europe that means Islam. At present, around 5 per cent of its population is Muslim. Immigration, combined with differences in birth rate, mean that by the middle of the present century the proportion will be four times greater (or rather less, if the Islamic decline in fertility continues).

That shift is trivial. The African explosion means that global Christianity is, for the time being at least, in robust health. As developed nations move away from the Church, half the people of Africa have joined. In 1900 there were around nine million believers there, but now it holds almost fifty times as many – a quarter of the world total. By 2025 there will be an even larger proportion, with seven hundred million, many among them followers of fundamentalist sects. Until the world's demography settles back into equilibrium, the faith that formed Europe will find its centre of gravity not in Rome, or in the Americas, but in mankind's native continent. A shared set of values has allowed one group, its ideas (and as an incidental its genes) to expand at the expense of all others. Like their equivalents in the Roman Empire, the descendants of those who join will boom in numbers not through biology, but through membership of a reproductively successful group. For them, if not for their competitors of other faiths or none, a common creed has been good news indeed.

That fact points at a great contradiction: that – like evolution itself – religions, as they grow and diversify, divide much more than they unite. As new sects emerge they are, like plants and animals, forced to compete for resources, or to disappear. Peace, brotherhood and moral principles then tend to fly out of the window. Christianity has had ten thousand denominations, many of which despised their rivals. The followers of the Catholic Eastern Rite still look down on their Western brethren and the Churches of Alexandria and of Antioch share mutual dislike. The Mormons have in their brief history splintered into seventy factions, from the Church of the Firstborn of the Fullness of Times to that of the School of the Prophets. Even the Church of God with Signs Following faces a schism between the brethren who handle snakes to demonstrate their trust in the Lord and those who drink poison instead. A recent outbreak of forcible beard-cutting among sects of the Old Order Amish has the same roots.

All this points at the central problem of faith: that while membership of a particular creed may be good for those who join, quarrels among them lead to disaster; as David Hume put it: 'Generally speaking, the errors in religion are dangerous; those in philosophy only ridiculous.' Christians, Muslims, Hindus, and Buddhists have been at sporadic war for their entire shared history, and splits within Christianity itself have been almost as bloody; in Byron's ingenious rhyme: 'Christians have burnt each other, quite persuaded/That all the Apostles would have done as they did.' Hatred starts young. Even in my Merseyside youth it was an automatic among my classmates to spit if they saw a nun and although such behaviour was no more than rude it was a forerunner to the sectarian clashes that began in Belfast a decade later.

Societies with the strongest devotion to their own truth and with the most expensive membership tickets tend to have the most conflict. Language is a dependable measure of division (as in the Book of Judges' account of a battle between splinter groups of Jews: 'Then said they unto him, Say now Shibboleth: and he said Sibboleth: for he could not frame to pronounce it right. Then they took him, and slew him at the passages of Jordan: and there fell at that time of the Ephraimites forty and two thousand'). Papua New Guinea once had dozens of native cults, some animist, some spiritual, and some downright cannibalistic. They were riven not only by belief, but by painful statements of sectarian identity such as deep incisions over the whole body. Neighbouring tribes now have more biological and linguistic differences than anywhere else, as evidence that belief has been a potent agent of exclusion and war. Matters are made worse by today's global fit of levels of devotion with poverty and insecurity. These too lead to discord; and the greater the power of the priesthood the happier the members of their flock are to die, or to kill, on command.

From the Fall of Jericho to the invasion of Iraq there have been some two thousand wars. Most were over resources or territory and were resolved by conquest or diplomacy. Blind dogma is more resistant to logic. About a hundred and twenty of those conflicts had a primarily doctrinal basis. They tended to last; the Wars of Religion – the sporadic battles between supporters of the Reformation and of the Catholic Church – continued for more than a century and some of its individual struggles such as the Dutch Eighty Years War and Central Europe's Thirty Years War were among the longest ever recorded. The Crusades grumbled on for even longer. Secular wars, in contrast, continue on average for less than half a decade.

More recent conflicts in the Balkans, the Middle East, Kashmir, Sudan, Sri Lanka, Iran, Iraq and the Caucasus have been ignited by such passions. Since 1945 there have been more than twenty wars between states, and more than two hundred clashes within them, with five million deaths in battle and five times as many due to disease and famine. More than half were ethnic or doctrinal in origin (while none, in spite of Marx, were caused by class conflict). For civil wars, like those between nations, there was a striking fit between how long they lasted and how ethnically (and often religiously) divided the nation had become. As Blaise Pascal put it, 'Men never do evil so completely and cheerfully as when they do it from a religious conviction', and he was right.

George Price would have been saddened to be reminded of that. He killed himself because he was ill and depressed and saw a lack of Christian charity all around him. He was buried in an unmarked grave, in a simple ceremony attended by just two colleagues and a few of his down-and-out friends and was almost forgotten until the importance of his research was realised, twenty years later.

Illness, depression and greed are still with us and mathematical biology will not cure them, but the story of George Price is both a tale of a scientific original who changed the course of modern biology and a reminder that – as he himself came to think – human society is as open to objective analysis as is that of any other creature.

Many of this volume's readers, I imagine, are happy to celebrate the decline of their nation's piety. The habit is an old one. Since the Enlightenment the imminent death of the Messiah has been hailed again and again, with almost as much fervour as his expected Second Coming. Even the devout have proclaimed his fall. William Blake himself, in his own interpretation of the biblical message, hints that God has already been ejected from the throne. His sins and those of Albion – the primeval man – meant that the deity had been transformed first into a mere mortal and then into Satan: 'That Human Form You call Divine is but a Worm seventy inches long.' Newton – science – was to blame ('A mighty Spirit leap'd from the land of Albion,/Nam'd Newton: he seiz'd the trump & blow'd the enormous blast!/Yellow as leaves of Autumn, the myriads of Angelic hosts/Fell thro' the wintry skies seeking their graves,/Rattling their hollow bones in howling and lamentation'). Blake felt that the only hope of salvation lay in God's Son, who after his crucifixion by the descendants of Newton on the Tree of the Knowledge of Good and Evil would be resurrected and would build Jerusalem in England itself.

In the secular sense at least, Blake's vision of a world reborn has been realised. In modern states, every taxpayer follows St

John's precept of 'unto others' whether they realise it or not. Finland and Sweden are among the least observant of nations, but their societies work far better when dealing with crime, illness and the like than do those of more pious countries. In modern times the most generous part of any social system is the State, and the efforts of the devout are marginal at best. Why, ask the sceptical, do people still need an excuse to be good? A divine edict, imagined or otherwise, was once a useful reminder that altruism works, but in the modern world surely it is not needed. Whether religion was invented as a means of social control or as an attempt to increase stability now means little, for in the West at least God's work has been replaced by that of Man. The decline of faith shows how peace, contentment and prosperity have come to depend more on human actions than on those of some imagined deity.

Over the years many dogmas from animism to Scientology have succeeded each other. Most have left little trace of their passing, and the same will no doubt be true of those practised today. The lesson from history is that in these connected times humankind could form a society that stretches beyond its own mental neighbourhood to embrace the globe into a single system of shared values. As the obstacles of speech, race and distance that once divided us are overcome, the time has come to abandon the last great restraint, William Blake's 'mind-forg'd manacles' of organised religion, which does so much more to divide than to unite. When those shackles are at last struck from their wrists, men and women, wherever they are, will no longer depend on the dubious promises of a serpent.

Instead they will be free to form a single community united by an objective and unambiguous culture whose logic, language and practices are permanent and universal. It is called science.

INDEX